AN AGRICULTURAL ATLAS OF ENGLAND AND WALES

J. T. COPPOCK, M.A., Ph.D.

With a Foreword by Sir Frank Engeldow, C.M.G., F.R.S., and a Technical Appendix by J. McG. Hotson, M.A.

'To one who cherished the illusion that he knew his English and Welsh farming scene intimately J. T. Coppock's *Agricultural Atlas of England and Wales* is an eye-opener; every page has its revelations . . . An essential reference book for every student, amateur and professional, of our contemporary agriculture.' *Countryman*

'Whether the student's approach is economic, social, geographic, climatic, purely practical or just statistical . . . the study of these maps and statistics provides all the answers a reasonable man might need.' *Guardian*

'. . . an impressive and valuable collection of statistics on the distribution of crops and livestock. It has about 200 maps but is much more than an atlas, with many pages of comment, explanation and analysis. A most important work of reference.' *Birmingham Post*

For this second edition of *An Agricultural Atlas*, which was first published in 1964, J. T. Coppock, Ogilvie Professor of Geography at the University of Edinburgh, has completely revised both the text and the maps, and has provided an additional twenty-nine maps.

A new and revised edition

THE CHANGING USE OF LAND IN BRITAIN

Robin H. Best and J. T. Coppock

'. . . the focus of their book is land use in all its aspects. They range from general questions about land use in Britain to particular studies, such as the changing pattern of land use in the Chilterns and the urbanisation of Tees-side. The book is both objective and informative, and should be compulsive reading for anyone who wants to talk about the loss of agricultural land, the density of urban development, the changing face of Britain and kindred subjects.' *The Economist*

'This is an excellent book for the planner, the agricultural specialist and the student.' *Municipal Journal*

hard covers

D1209424

An Agricultural Atlas
of England and Wales

THE CHANGING USE OF LAND IN BRITAIN
by Robin H. Best and J. T. Coppock

AN
AGRICULTURAL
ATLAS OF ENGLAND
AND WALES

by

J. T. COPPOCK, M.A., Ph.D.

Ogilvie Professor of Geography
University of Edinburgh

With a Foreword by
SIR FRANK ENGELDOW, C.M.G., F.R.S.
Emeritus Professor of Agriculture
University of Cambridge

and a Technical Appendix by
J. McG. HOTSON, M.A.
Department of Geography
University of Edinburgh

FABER AND FABER LTD

3 Queen Square

London

First published in 1964
by Faber and Faber Limited
3 Queen Square London WC1
New and revised edition 1976
Printed in Great Britain at
the Alden Press, Oxford

ISBN 0 571 04829 3

Contents

Maps[1]

[1] Agricultural maps refer to June 1970 unless otherwise stated.

11

15

Maps

Acknowledgements

This book is an attempt to provide a view of those features of the agriculture of England and Wales for which convenient, mappable data exist and is based largely upon the district summaries of the annual agricultural returns. I have received much help and advice during its preparation and it is impossible to acknowledge by name all those who have answered my queries. I owe a special debt to the following people: Dr. C. Board (University of Cambridge), Mr. G. Allanson and Dr. R. Best (Wye College, University of London), Mr. A. Hay (Association of Agriculture), Mr. K. Hunt (Agricultural Economics Research Institute, University of Oxford), and Dr. B. E. Cracknell, Mr. E. Snowdon and Miss O. G. Tanner (Ministry of Agriculture, Fisheries and Food), who commented on the original proposal; Mr. A. N. Croxford, Mr. R. V. Knibbs and Miss D. Dixson (Ministry of Agriculture, Fisheries and Food), who made available most of the original material on which this atlas is based; the British Sugar Corporation, the Fruit and Vegetable Canning and Quick Freezing Research Station, the Milk Marketing Board and the Potato Marketing Board, who provided additional information; Dr. E. H. Brown (University College London), and Dr. D. A. Osmond (Soil Survey of England and Wales) who allowed me to use unpublished maps; Dr. A. Booth, Dr. M. Levison and Mr. A. Sentance (Department of Numerical Automation, Birkbeck College, University of London), who made it possible for the calculations to be undertaken by computer; the Director of the Computer Unit, University of London, for access to the Mercury Computer; Miss J. Morris, Messrs. J. Bryant, K. Wass and M. Young (University College London), without whose cartographic skill in preparing my drafts for publication the atlas would probably never have been produced; to Messrs. G. Allanson, R. Millbourn and R. R. Folley (Wye College), Professor W. R. Mead, Mr. D. Thomas and Dr. G. Ward (University College London), Dr. C. Board and Mr. B. Jackson (University of Cambridge), Professor D. K. Britton and Mr. G. E. Jones (University of Nottingham), and Dr. B. E. Cracknell, Mr. P. G. Ellis, Mr. D. D. Filtness, Mr. E. Snowdon and Miss O. G. Tanner (Ministry of Agriculture, Fisheries and Food), who have read and commented on parts of the manuscript. I am particularly grateful to Professor Sir Frank Engeldow, who not only stimulated my original interest in agriculture when I was a post-graduate student at Cambridge, but has also made many helpful comments on the atlas and has kindly written a foreword to it. I have also been much helped by the friendly assistance I have received from the publishers; thanks are especially due to Mr. Morley Kennerley, particularly for his

17

sympathetic understanding when the project had temporarily to be abandoned as a result of my father's illness and death, and to Mr. M. Shaw, who has guided my footsteps through the intricacies of book production. I am grateful to my wife who has helped with the checking and has applied her classical training to the task of removing obscurities and ambiguities in the text; she has also kindly prepared the index. Lastly, my thanks are due to Professor H. C. Darby, in whose department the work was conceived and executed, for providing the facilities without which the atlas could not have been produced. To all these and many others I am most grateful; but the responsibility for any errors or omissions remains mine alone.

J. T. Coppock

University College London, 1964

Author's Preface to the Second Edition

The second edition of this atlas presents a snapshot of agriculture in England and Wales in 1970. In preparing it, I have kept the original format, making such alterations as were necessary because items were no longer separately recorded in the census, or because new items had been included and warranted a place, or because the text was no longer correct. Had the atlas been recast, there could have been no justification for the prominence given to oats, mixed corn, turnips and mangolds, though the agricultural census remains much more informative on crops than on livestock, despite the importance of the latter in the economy of the great majority of farms. I have also retained the preface with which Sir Frank Engeldow graced the first edition; for it was under his stimulus that my initial interest in agricultural distributions was strengthened.

The major change in this edition is in the maps themselves, which, apart from those few retained from the first edition, have nearly all been prepared by computer and printed on a modified line-printer. How successful these are will be for readers to judge, but without the computer's aid a second edition of this atlas would not have been possible. I am grateful to my colleague Jack Hotson (who contributes a technical appendix), Eric Anderson, Alison Finch, Margaret Wilkie, Richard Gillanders, and others who have all helped to make these maps, although any errors, and I am sure they exist, are mine. Some have come to light only as I tried to match text to map, but the most difficult errors to detect are those which are most plausible.

This edition is a by-product of a research project, conducted in collaboration with the Ministry of Agriculture, Fisheries and Food and the Department of Agriculture and Fisheries for Scotland, and financed by the Social Science Research Council, into types of farming in Great Britain. I am particularly grateful to Mr. P. Horscroft and his colleagues at the Agricultural Census Branch of the Ministry of Agriculture for supplying the information on which the maps are based and answering my numerous queries.

Since this revision was begun, the United Kingdom has joined the European

Acknowledgements

Economic Community. This atlas describes the agricultural scene before the effects of this great venture have begun to be felt. It will be instructive to compare these maps with those that might be prepared for a third edition a decade hence.

J. T. Coppock

University of Edinburgh, 1974

Foreword

Up to the early years of this century teaching in agriculture had two almost independent parts, broadly speaking, practice and science. Customary types of farming and husbandry practices were described, often in detail, for the chief geographical regions and classes of crop, grass and livestock of the country. For the sciences the applications of botany, chemistry and to a lesser extent animal physiology were taught together with some economics, book-keeping and law. Factual description, with little interest in rationalization, was the basis. A correspondingly limited outlook prevailed in research. World War I, involving a food crisis, induced in the belligerent countries a more fundamental interest in the use of land. In the ensuing quarter of a century a number of happenings strongly fortified this. Colonial responsibilities gave many young agriculturists, practical and scientific, the rousing experience of strange crops, climates and agricultural systems. Soil erosion at last began to be recognized as a grave menace. And the universal financial crisis of the late 1920s stirred up a keener interest in the business side of farming. World War II had a far stronger effect and also made everyone aware of the world's food-and-population problem.

By these influences the teaching of agriculture was lifted to a new level. In the parent subject—husbandry—factual description was preceded by presentation of the principles deriving from those climatic, soil, biological, economic and social factors which everywhere determine the agricultural use and possibilities of the land. Illustration of these principles constituted a study of comparative agriculture at large, and, more intensively, of the country.

Advances in soil science encouraged the new outlook. Climatic influence on soil formation, international soil classification and mapping of soil types opened an important field for geographers. Their joint work with ecologists and agriculturists began to bring out the relationship between the world's climatic regions and systems of agriculture. At the national scale, the maps and reports of the Land Utilisation Survey drew attention to many such relationships as they existed in the 1930s.

Contemporaneously agricultural economics in Britain and elsewhere gained new life from the analysis of systematically collected primary data. The annual *Agricultural Statistics* (first issued in 1866) were increasingly supplemented by quantitative information. This, among other advantages, helped to portray the farming of the main regions of the country. With the setting up of the National Agricultural Advisory

Service and of the Provincial Agricultural Economic Service in 1947 the country became strongly equipped for the study of the composition, practices and policy of its agriculture.

By degrees, economics has acquired an important place in agricultural teaching, both institutional and advisory. It has enabled practical husbandry to be presented in quantitative economic terms as well as technically. Not less importantly, it has promoted the interest of students, teachers, research workers, farmers, administrators and governments in the national aspects of agriculture. In this matter national maps have been essential. The *Agricultural Atlas* published in 1925 and the Ministry of Agriculture *Map of Types of Farming* (1941) have served well. In this new atlas Dr. J. T. Coppock offers a very comprehensive display of the many aspects of the farming of England and Wales. What guided this choice of several possible bases and forms of presentation is carefully set out in Chapter I. In this the limitations and uncertainties arising from the form and scope of available information are made plain. This chapter affords a necessary and valuable commentary for those unfamiliar with the intricacies and pitfalls of agricultural statistics.

The author's aim has been to display the physical and economic background of the country's agriculture by maps and an explanatory commentary on what they show. In any study in agriculture, whether general or highly specific, whether experimental, economic or geographic, it is at the least interesting and often essential to examine the relevant national aspects. For teachers and researchers in several disciplines together with those farmers, administrators and others who take a wide interest in the national agriculture this atlas provides an abundance of information and a clear presentation.

F. L. ENGELDOW

CHAPTER I

Aims, Sources and Methods

Even to a casual observer it is obvious that there are marked differences in agricultural activity throughout England and Wales; thus Kent and Worcester are associated with fruit growing, the eastern counties with arable farming, Cheshire and Somerset with dairying, and the Welsh uplands and the Pennines with sheep rearing. There are also many minor differences in the crops grown and the livestock kept which help to distinguish one area from another, such as the rhubarb growing of the Leeds–Wakefield area or the turkey farming of Norfolk. To get any countrywide view of these differences and to give precision to our vague impressions it is necessary to rely on the statistical data collected by the Ministry of Agriculture, Fisheries and Food (hereafter referred to as the Ministry of Agriculture) and by other official bodies. Yet these data cannot easily be assimilated as tables of statistics; they need first to be mapped.

The object of this book is, therefore, to present some features of the agriculture of England and Wales in maps. What is shown is limited by the information available; the book cannot aim to be exhaustive and many aspects are necessarily omitted, not from choice, but because no mappable data exist. Nor is all the information mapped of equal importance. Many minor crops which play similar roles in the farm economy have been included because of interesting features of their distribution. On the other hand, although livestock are the most important element in British farming, the book is biased in favour of crops because the crop data are less ambiguous. The accompanying text is intended primarily to clarify and comment on the maps and, where possible, to provide explanations of what they show.

The book is divided into nine chapters, of which the first is an introduction to the sources and methods. The second and third contain maps, often from sources other than the agricultural returns, showing the physical basis of agriculture and some of the relevant economic considerations. The fourth and fifth chapters are devoted to maps of tillage crops and grass, the sixth to horticultural crops. The seventh chapter is concerned with livestock and shows both the distribution and the relative importance of the main classes of livestock. In the eighth chapter, the most speculative of all, an attempt is made to measure the intensiveness of farming and to show how the different farming enterprises are combined in each district, while the epilogue discusses future research. Appendices I and II contain further discussion of the ways in which the maps were produced, while Appendix III contains details of the various crop, livestock and enterprising combinations, and some selected statistics.

23

Data

Most of the maps are based upon data from the agricultural census of 4 June 1970. This is one of a series of censuses which have been made annually since 1866 by collecting returns from occupiers of agricultural land. These individual returns are consolidated into parish summaries, which in turn are grouped to give county and national totals. Since the details of individual returns may not be disclosed, it is necessary to use these summaries for administrative areas in preparing maps. Most of the maps are based upon the 380 districts of the Agricultural Development and Advisory Service (A.D.A.S.). These districts are subdivisions of counties and generally comprise groups of some 30 or 40 parishes; their boundaries, as they existed in 1970, are shown in Figure 12. For data which are not available by districts the county has generally been the mapping unit.

The year 1970 has been chosen for three reasons. First, it was the most recent year for which data were available in convenient form. Secondly, it coincided with a decennial World Agricultural Census. Thirdly, it seems reasonably representative of agriculture on the eve of British entry into the European Economic Community and so provides a bench-mark against which subsequent changes in agricultural distributions can be measured. No one year can, of course, be truly representative. On top of long-term trends, such as the gradual replacement of the horse by the tractor, there are annual fluctuations in the acreages of crops and in the numbers of livestock. These are due in part to differences in the sequence of weather from year to year which affect the ease of autumn ploughing, the availability of feed, the success or otherwise of the harvest and, in the uplands, even the mortality rate among livestock. Other annual fluctuations are due to the adjustments by farmers to changes in agricultural prices. How exactly these operate is a matter of some debate. Largely because of changes in prices received by farmers, most livestock show cyclical variations in numbers which are most marked in the case of pigs. Fluctuations in crop acreages are less regular and adjustments to price changes seem generally slower, particularly in the case of fodder crops. A classic example of changes due to adverse weather is provided by the severe winter of early 1947. Between the census of June 1946 and that of June 1947 total sheep numbers fell by over $2\frac{1}{2}$ million and the breeding flock by over 1 million, and it required 5 years for numbers to be restored to their previous level. An example of changes due to variations in prices is the fall in the number of pigs by 9% between 1958 and 1959, following the reduction in guaranteed prices for pigs in the 1958 Annual Price Review.

In view of these fluctuations it might have been preferable to have used average values for 2 or 3 years. In practice, this was not possible. No up-to-date map of districts was available and the present map had to be laboriously constructed from parish lists. Moreover, there are changes in the composition of districts, so that district totals in successive years are not strictly comparable. Nevertheless, while the absolute differences between districts vary from year to year, the relative importance of districts remains, in the short-term, fairly constant. Furthermore, while no year can be called normal in view of the variability of the British climate, the acreages of crops and numbers of livestock recorded in June 1970 do not seem atypical.

What can be mapped is largely limited by the questions asked in the June census. Even so, difficulties of interpretation inevitably arise, for it is impossible to fit the variety of British farming to the Procrustean bed of uniformity demanded for statistical purposes. As far as crops are concerned, there are, with the exception of certain vegetables, no serious problems in determining the acreages under the various crops, for most crops are in the ground at the time of the census and errors in the measurement of areas are unlikely to affect their relative importance. The vegetable acreage is certainly understated in June, for some areas are double-cropped, while vegetables such as spring greens are largely cleared before the census and others, such as turnips and swedes, have not all been planted by that date. Catch cropping, with crops such as mustard, rape and ryegrass, may also escape enumeration in the June census. Another minor difficulty is that, with the exception of mixed corn, mixtures of crops are not recorded as such, but are apportioned among their constituent crops. Again, some crops are only partially enumerated under a separate heading, e.g., only rye for threshing is separately shown, the small acreage of rye cut green being included under 'other crops'. For some crops the purposes for which they are grown are embodied in the definition used, e.g., beans for stock feeding, but this is not always done, partly because the answer cannot accurately be known until the crop is disposed of, e.g., no distinction can be made between barley grown for feeding and barley grown for malting, for many farmers hope to produce a crop of malting quality.

Because no district data on yields are available, acreage and stocking density respectively must be the chief criteria in estimating the importance of different crops and livestock. Nevertheless, there are considerable differences in yields of crops and livestock products throughout the country and some allowances ought to be made for these. Nor is it possible to provide maps from census data to show the distribution of different varieties of crops and breeds of livestock, although many of these have distinctive distributions and ecological preferences.

Major difficulties arise over the interpretation of different kinds of grass, chiefly because there are no clear-cut dividing lines between them and because practices vary from one part of the country to another; thus in western areas, temporary grass in long leys grades imperceptibly into periodically reseeded permanent pasture, while in East Anglia, where 1-year leys are common and much permanent grass is unploughable, the distinction between temporary and permanent grass is much clearer. Inclusion of questions about the age of leys, by clarifying what should be returned as temporary, has been partly responsible for an increase in the acreage returned as temporary grass. In view of differences in the quality and composition of grass throughout the country, the rough grazing/permanent grass boundary is even more arbitrary, for it is very difficult to ensure that poor grass is everywhere returned in the same category; some poor permanent pasture in the uplands would probably be returned as rough grazing by a lowland farmer, while there has probably been considerable upgrading of rough grazing to permanent pasture in recent years.

The data relating to numbers of livestock are generally more difficult to interpret than those relating to crops and grass, partly because age rather than type is often used to distinguish different classes of livestock. There is, for example, no separate class of female cattle being fattened for beef, but only a category, female cattle 2 years old and over, which includes both dairy herd replacements and old cows intended for slaughter. In view of the current importance of beef from dairy herds, this limitation is a serious handicap. An even more important difficulty arises from the fact that livestock numbers are not the same throughout the year. 4 June is a less satisfactory date for the collection of livestock numbers than it is for the collection of crop acreages. The cycle of animal life has a different seasonal pattern from that of crops and varies throughout the country and between different classes of livestock; thus lambing is spread over several months from December in the lowlands to late spring in the uplands, and, while cows may calve throughout the year, spring and autumn calvings are most common, though even here there are regional differences. More important is the fact that the distribution of livestock, particularly sheep and beef cattle, varies from season to season, as animals are moved between farms or from farm to slaughterhouse. In general, there is an eastward movement from the rearing areas in the hill country of the west and north, but the pattern is exceedingly complex. Beef cattle may pass their lives on several farms in different parts of the country, being bred on one farm, reared on another and fattened on a third; they may also be imported as store animals from Ireland or Scotland. Ewe lambs, born in the hills, may winter in the lowlands, spend several seasons breeding on the upland rough grazings, then migrate to more distant lowlands where they rear one or more crops of lambs before being slaughtered. In Kent, the leading lowland county for sheep, a similar exchange takes place between Romney Marsh and the surrounding 'upland', with sheep wintering away from the exposed marsh, and stocking densities may differ considerably in summer and winter. Although quarterly censuses are taken in September, December and March, they are sample censuses relating to one-third of the holdings and no district data are available, while county estimates may be subject to error, particularly for the smaller counties. Furthermore, movements of livestock, particularly on the hill margins, take place within the boundaries of a single county, so that a county map gives only a crude picture of seasonal differences. Fortunately, the movements of dairy cattle, the principal class of livestock on British farms, are less important than those for other classes of stock, for many farmers rear their own replacements, and seasonal differences either in numbers or in their distribution throughout England and Wales are small.

It would be illuminating to have maps not only of the distribution of stock at different times of the year, but also of the flow of livestock from one part of the country to another. It would also be desirable to have flow maps of the disposal of crops and livestock products, particularly in view of the localization of production and the varying distances separating producing and consuming areas. Such a map was produced in the report *The Fluid Milk Market in England and Wales* (H.M.S.O., 1927). An up-to-date map would be invaluable, as would one showing the movement of horticultural produce, particularly in view of G. R. Allen's suggestion that a considerable part of home-produced fruit and vegetables is unnecessarily routed through London markets. Unfortunately the data necessary to produce such maps do not exist.

Many problems of interpretation would arise even if individual returns could be mapped, but the use of parish, district or county summaries presents further difficulties. The district data are the amalgamation of several hundreds of returns from individual farms. The degree to which these farms differ from each other varies greatly from one part of the country to another depending on both physical conditions and farm organization; it is likely to be least where physical conditions are adverse and only a limited range of farming activities is possible, as in the Welsh mountains, and to be greatest in areas where climate and soils permit a wide range of crops to be grown and many kinds of livestock to be kept. These differences are well illustrated in R. Bennett Jones' study, *The Pattern of Farming in the East Midlands* (1954); thus, while in Nottinghamshire no one type of farming accounted for as much as 20% of all full-time farms and the different types were widely distributed, dairy farms accounted for over 75% of all farms in Derbyshire and were virtually the only type of holding over large parts of that county. Maps based on the district summaries cannot show the extent to which individual farms depart from the mean values which the district figures represent, although where a single enterprise accounts for a large proportion of the farming activity in a district, the possibility of variation from farm to farm can only be small. It must be appreciated that by grouping farms of different kinds, the district data may minimize the extent of local specialization in particular crops or stock. Without access to the individual returns it is not possible to say how far the distinctive farming activities of a district occur on all farms or in only part of the district or only on holdings of a particular size.

While county, district or parish data could be used for the preparation of such maps, the district has been chosen for several practical reasons. It is clearly superior to the county, being more homogeneous in physical characteristics and more uniform in size. For whereas the counties range from nearly 2,800 square miles in the West Riding to 83 square miles in the Soke of Peterborough, the largest district is some 400 square miles and the smallest 15. Even this range is too large, for districts of different sizes represent varying degrees of generalization of farm data and these may produce differences on the map which are only apparent, i.e., the distribution may appear more uniform where districts are large, and more fragmented where they are small. Consequently, some of the smaller districts have been amalgamated.

It is true that the parish would be a superior unit, but there are more than 12,000 parishes in England and Wales, and the cost and effort involved in using them would not have been justified by any benefit that would have been gained. At the scale at

Aims, Sources and Methods

which the maps have been reproduced, the additional detail would have made for loss of clarity, as Figures 15 and 16, which are based on parishes, suggest. Furthermore, although the parish is generally considered smaller than the district, it is often not markedly more homogeneous, particularly in scarpland and hill country, so that the margin of advantage in using the parish is not as great as it appears at first sight.

Nevertheless, the suitability of the district as a mapping unit varies considerably. Some districts are broadly coextensive with minor physical regions; for example, District 32 corresponds approximately to Holderness, District 69 to the Fylde, and District 237 to Romney Marsh. Others, particularly those in scarpland country, seem to have been designed to embrace as many types of land as possible, e.g., District 231, which includes parts both of the North Downs and of the Weald. The degree to which districts and physical regions correspond can be established by comparing Figures 11 and 12. In some counties, such as Oxfordshire, there is little correspondence, but in others, such as Norfolk, most of the minor divisions of the county—the Breckland, the Fens, the Broads, the Good Sands region—can be equated with particular districts or groups of districts, though even here natural and administrative boundaries do not coincide exactly, so that regional differences are often somewhat muted.

The Maps

Most of the maps are choropleth maps, in which regional differences in the importance of particular crops and livestock are shown by differences in the density of shading. While there are many ways of showing statistical material of this kind, choropleth maps seem the most satisfactory method on balance. Numerous divided circles are visually unsatisfactory and the only effective choice is between dot maps and choropleth maps. It must be admitted that the dot maps often look more pleasing and that no arbitrary decisions about class boundaries are involved. Yet the dot map, by omitting administrative boundaries, gives the impression of being more precise than it is, particularly if the attempt is made to distribute dots in accordance with preconceived ideas about their probable distribution; for no cartographic device can alter the fact that the data are generalizations which cannot be specifically related to any part of the area to which they refer. It is also difficult to find a satisfactory value for each dot where there is a wide range of densities, for at low densities, values may be too low to be shown, while at high densities the dots fuse together. Again, the impression of distribution which dot maps convey can only be very imprecise, and it seems a pity to neglect the possibilities which quantitative data offer of giving some more adequate measure of differences. In any case, dot maps can be used only to show distributions; some other method must be employed to show ratios and similar relationships.

Given the nature of the data and of the districts, the choropleth map has a number of advantages. It provides a quantitative measure of the distribution which is easily grasped; it links together areas which have certain features in common; and it clearly shows that it is based on administrative areas. It is true that it has several corresponding disadvantages. For practical reasons only a small number of shadings can be used; the eye cannot easily discriminate between many different shadings, while the larger the number of classes the more fragmented the resulting map and the more difficult it is to read. This limitation to five or six shadings may not be wholly disadvantageous, for many of the small differences between adjacent districts are probably unimportant.

A more serious disadvantage arises from the necessity of choosing values which particular shadings are to represent. The maps are intended chiefly to show the regional differences between one part of the country and another. It is impossible to choose class intervals which are satisfactory for all areas, and the location of some boundaries and the exact shape of some of the shaded areas is therefore fortuitous. Although it is quite clear which are the most important and which the least important areas, quite small differences can place adjacent districts in different categories, especially where the distribution being mapped is widespread and the range of values between the least and most important areas is small. While many of the shapes of shaded areas, particularly those with shadings representing high values, are relevant, all that can be said with certainty about a specific district is that the acreage of crops or number of livestock lies somewhere within the range of values defining the particular category in which it has been placed. The choice of class interval is necessarily somewhat arbitrary, for it is a compromise between two ideals, the selection of values which group areas with similar characteristics, and the production of a map which is easily comprehended. Intervals ought, as far as possible, to correspond to major gaps in the frequency distribution of different values; on the other hand, regular intervals are most readily understood. The mappability of the chosen classes is also important, i.e., the extent to which they produce homogeneous blocks of shading, rather than fragmented patterns, although this will largely depend on the nature of the crop or class of livestock. These aims cannot all be satisfied, though they have all been taken into account in choosing limiting values. Most intervals are in multiples of 5 or 10, but different intervals are employed on a few maps and, on others where the crop or livestock is highly localized, the values of the intervals increase in geometrical progression, since it was thought desirable to give some indication of differences within the low value areas. While it would have been preferable to choose the same intervals for all maps, since this would make comparison of the maps easier, this is clearly impossible, and the key to each map must be carefully studied, since the same shading represents different values on each map. Most crops and grassland have been shown as proportions of the acreage under crops and grass, and care must be exercised in interpreting results for those districts where there are large acreages of rough grazing. Densities of livestock have generally been calculated in relation to all agricultural land, i.e., crops, grass and rough grazings.

Although dot maps of a kind can be produced by line-printer, only choropleth and isopleth maps can satisfactorily be produced by this means. In view of the small scale of the maps, isopleth maps were not considered suitable. For maps of horticultural crops, which are often highly localized, acreages have been shown as a proportion, not of tillage or crops and grass, but of the acreage under that crop.

The Text

The text, in addition to giving certain factual information about acreages and numbers, draws attention to the main features of the accompanying maps and outlines the changes in the preceding 32 years, and also attempts to suggest explanations where this is possible. It is in one sense easy and in another extremely difficult to explain the regional differences which these maps show. Many of the differences are of long standing and relate primarily to the physical geography of the country; for Britain is a long-settled country of great physical variety where trial and error over the centuries has often produced a close adjustment between man and the land. The pattern of soils is intricate and complex, while climatic conditions also vary markedly from place to place, not only in respect of the amount and incidence of rainfall, but also in temperature regimes and in length of growing season. Yet, while it is obvious that these physical differences favour some farming enterprises and make others difficult, it is not easy to demonstrate exactly how they operate. In large measure this difficulty is due to the absence of data about the effects of soil and climate on crop and livestock performance, a not surprising lack in view of the fact that yields of crops may vary quite markedly in different parts of the same field and under different management on the same terrain. Furthermore, crops and livestock do not exist in isolation. Not only do they compete with each other, but what is grown or kept is determined partly by the needs of the farming system as a whole; as Astor and Rowntree put it, 'it is almost true to say that in British agriculture nothing is ever done for its own sake alone'. Thus crops may be grown as part of a rotation, the aim of which is to keep the land in good heart, to keep down weeds and to control pests and diseases, and it has proved difficult to devise satisfactory rotations in which cereals do not play an important part, although the choice of cereal to be grown is determined by other considerations. On the other hand, crops like sugar-beet and potatoes cannot be grown continuously without regard for crop hygiene, however suitable the land; heavy eel-worm infestation has resulted from the too frequent growing of these crops, and rotations in areas such as the Fenland have had to be lengthened by the inclusion of other crops, e.g. clover and peas.

Economic considerations are also very important, for while the ecological requirements of crops and livestock set ultimate limits on what is done, the choice, in countries such as Britain where farming is primarily an economic activity, will be determined chiefly by profitability. Apart from price levels themselves, which are often uniform over large areas, many factors such as soil, climate, farm and field size, and location, affect the profitability of growing different crops or keeping different kinds of

30

livestock in each locality. Furthermore, it is not only the differences in the profitability of a particular enterprise which determine whether it is undertaken. Comparative advantage is even more important. Thus, while returns from dairying may be lower in the Exmoor fringe than in the Somerset Levels, cattle rearing may be even less profitable, so that many farmers keep dairy herds, although conditions are far from ideal. Under the Farm Management Survey a great deal of information has been gathered about the profitability of different enterprises on a relatively small sample of farms in which the larger farms are over-represented, and some information is available for the different regions of the country about the general level of profitability on these farms as a whole. But there are few countrywide studies of the profitability of different enterprises which show whether there are, in fact, any substantial differences in costs between the various regions; indeed, those which exist and which are based on sample data throw very little light on this question or produce results which seem quite implausible. It is generally necessary, therefore, to infer comparative advantage from what is known of the physical requirements of different crops and livestock, and from what most farmers do.

Such assumption must obviously be made with caution. Farming is a complex of operations in which it is often difficult for the farmer to know whether a particular activity is profitable or not; this is especially true of the growing of fodder crops which do not leave the farm. Furthermore, decisions about the choice of farming systems are made by a very large number of farmers who differ greatly in business ability, farming skill and in their knowledge of the profitability of different systems. For these reasons, few farmers knowingly achieve the maximum profit possible with their resources.

Farmers also vary considerably in the extent to which they are governed by non-economic motives. Tradition plays an important part in farming practice. The growth of some crops in particular areas can partly be explained by appeals to tradition, e.g., the Cornish preference for mixed corn as a fodder cereal. Long-established practices may be rational, but they may not, as with the chalking of land on the Yorkshire Wolds, which has been shown to confer no benefits. Again, the original reason for the establishment of a particular crop or livestock enterprise, though once valid, may have ceased to have any relevance. Thus pig keeping in Cheshire was formerly related to the production of farmhouse cheese and butter, but although these have virtually ceased and been replaced by the sale of liquid milk, pig keeping nevertheless persists as an important enterprise in the county.

Other non-economic attitudes may be relevant to an adequate explanation of the distribution of particular crops and livestock, e.g., the pleasure that many farmers get out of livestock, or the feeling that farming without livestock is not proper farming. One of the most interesting studies of attitudes of this kind is that made by J. B. Butler in *Profit and Purpose in Farming* (1960). He found that many differences in farming systems in a group of parishes in the Vale of York could be explained by factors such as age, temperament, education, and whether a farmer was native to the area or not. This last consideration has been a very important one, for farmers from other parts of the country have often brought new ideas and crops and acted as catalysts in bringing about changes in the regional type of farming. A classic example is provided by the migration of Ayrshire farmers to parts of the Home Counties in the late 19th century, bringing with them a knowledge of dairying, potato growing and ley farming. More recently, migrants from Lincolnshire have been instrumental in extending the area under cash crops in Romney Marsh. In so far as characteristics such as high average age and lack of formal education are regionally grouped they are likely to be relevant in explaining regional differences in agriculture. There is some truth in the dictum that good farmers tend to gravitate towards good land, and there is similarly abundant evidence that farmers on the hill margins are often elderly and conservative; thus some two-thirds of the farmers in an upland area surveyed in mid-Wales were over 50 years of age. Furthermore, many farmers are content to receive a lower income than they could earn as agricultural labourers, presumably because they value their independence and because farming is for them 'a way of life'.

Chance, too, plays an important part in explaining the origin of crop and livestock distributions, particularly of minor localized enterprises. An individual successfully introduces a new enterprise and his example is then followed by others. A striking instance is provided by asparagus growing in Breckland which was introduced by Lord Fisher of Kilverstone in 1933. Some years before he had noticed an asparagus plant growing wild on heathland; it flourished each year and this encouraged him to experiment with the crop. The area is now one of the two major centres of asparagus growing in the country. It is interesting to notice that it is an area of light soils, and that the crop is here grown on a large scale by a few farmers. These conditions contrast markedly with those in the Evesham area, formerly the main centre; here the crop is grown by smallholders on heavy loams.

While an area must provide at least minimum conditions for a crop, it must not be assumed that the localities in which a crop is grown represent either the only or the most suitable areas. Specialization, which may have arisen by chance, tends to be perpetuated by the development of ancillary services; thus the concentration of cherry growing in Kent has been helped by the existence of a skilled female labour force for the difficult task of picking. Similarly, different areas acquire a reputation which has a certain market value, e.g., Blackpool tomatoes or Lincolnshire potatoes. On the other hand, expansion of demand for a highly localized crop not infrequently leads to the extension of production on to less suitable terrain; thus market gardening in the Evesham area has spread from the river terraces where it originated on to the heavy Lias Clay. Conversely, there are undoubtedly suitable areas for the production of vegetables, especially earlies, which are handicapped by isolation or where the crop has simply never been tried.

The influence of such differences in accessibility is also difficult to evaluate. With perishable crops in which freshness is an important consideration, e.g., the production of lettuce, location near the fringe of a large urban market has obvious advantages. Bulky commodities, like sugar-beet, are similarly tied fairly closely to their markets, in this case, the sugar factories. Potatoes, on the other hand, are extensively grown in the Fenland, some 80 or more miles from the main urban markets, while there is regular movement of carrots from the Fenland to Scotland. Improvements in communications, rises in other costs, the development of complex systems of marketing in which contacts are more important than location, and the development of pricing policies which minimize the effects of distance, notably in the case of milk production, all modify the effect of location. Unfortunately little information is available about the disposal of produce and about costs of transport from producing to consuming areas.

Explanations offered must therefore be incomplete and, in the absence of suitable economic data, will tend to emphasize physical controls. It is hoped that the maps themselves will be a challenge to produce more adequate explanations.

Notes on the Maps

In the maps printed by line printer, the figures underneath each block in the keys indicate the upper limit of the class represented by each category of shading.

Thus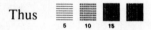

The first category is less than five, the second is between five and ten, the third is between ten and fifteen and the fourth greater than fifteen.

The scale of the full-page maps is the same as on Figure 1 and that of the quarter-page maps as on Figure 2.

CHAPTER II

The Physical Basis of Farming

Agricultural activities are governed in large measure by relief, soil and climate. Unlike knowledge of the man-made conditions of farming, which is often very imprecise, information about many aspects of physical geography is generally available in considerably greater detail than the agricultural data shown in subsequent sections, although the way in which these physical factors operate has been little studied. In this chapter the main physical characteristics of England and Wales, as they affect agriculture, will be briefly examined to provide a perspective for a consideration of the various crops and livestock in later chapters.

Relief

Relief influences agriculture in two principal ways; by affecting the ease with which the ground can be cultivated and by modifying climate (Fig. 1).

In most of Lowland Britain differences in relief are less important than soil differences. It is true that height does affect the length of the growing season, crops on the escarpments being a week or more later than those on the adjacent lowlands. Aspect and slope affect earliness and frost incidence and are particularly important in the cultivation of fruit and vegetables. While flat or gently undulating terrain is most suitable for mechanized cultivation, few areas are too steep to be ploughed, the scarps of the Lower Greensand, the Chalk and the Jurassic limestones and sandstones being the chief examples. Some land is not ploughed because it is too low-lying or because of a high water-table. The area liable to flooding is comparatively small, being largely confined to the major river valleys and the coastal marshes, while only the peats of the Fenland lie below sea level. High water-tables are locally a problem, although in the Fenland the water-table is maintained artificially several feet below the surface. Yet the chief reason why much low-lying land is not ploughed is to be found in the prevalence of heavy soils rather than in relief.

In the uplands of the west and north relief plays a much more important part. These uplands consist essentially of a series of dissected plateaux; the residual surfaces are flat or gently sloping, and often suffer from poor drainage, while many of the dissecting valleys have steep sides which are unploughable and are often under wood or bracken. Although many of the gentle plateau slopes are ploughable, their poor drainage and the rapid change of climate with elevation, especially the shortened

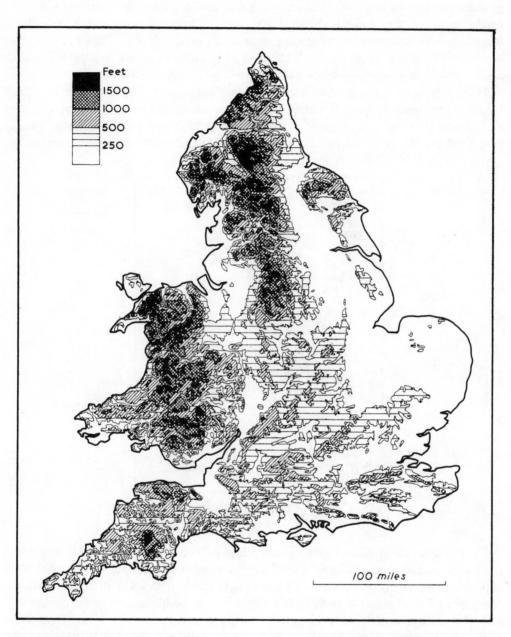

Fig. 1. Relief

growing season and increased humidity, are major obstacles to their cultivation. As a result, most of the upland above 1,000 feet is covered with rough moorland. Ruggedness as such is a handicap only over small areas, chiefly in the Lake District and North Wales, most of which are too high to be cultivated; in a few localities, such as the Craven limestone uplands, bare rock surfaces prevent cultivation.

There is thus an upper limit of improved land, although it does not occur at a fixed height everywhere. In part it varies with economic conditions, falling in times of agricultural depression and rising when prices are high and costs relatively low. In part it reflects the long history of agricultural colonization of the upland margins, for

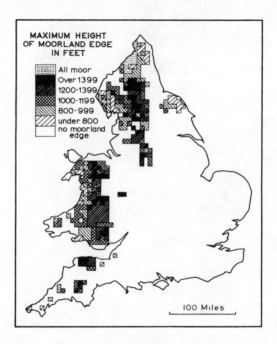

Fig. 2. Height of the moorland edge.

this was the work of generations of individual pioneers, and many local anomalies mark past successes and failures in land improvement. In the broad view, the height of improved land varies with the size of the upland mass, with its latitude and with the oceanicity of its climate; it tends to fall northward, to be highest on extensive uplands and to rise from west to east. In detail, rock type and the nature of the terrain are important. Figure 2 is an attempt to give some precision to general impressions and shows the height of the boundary between moorland and continuous stretches of improved land for each 10-kilometre grid square. The highest limits are found in the sheltered interior dales of the central Pennines; the lowest limits are in west Wales, the Lake District and the smaller uplands of the south-west.

Climate

Despite the small area of England and Wales and the limited range of elevation, climatic differences are very important in accounting for differences in the kind of agriculture practised. The ways in which they do this have not been extensively investigated and it is probable that the operation of some controls is more subtle than it appears, for crops can be affected by conditions at germination, during growth and at harvesting. Livestock are affected by the extent to which they need protection from the weather and, indirectly, by the supply of feeding stuffs. Controls may also act indirectly by their influence on the viability of pests and diseases.

The amount and frequency of rainfall are important chiefly in determining whether conditions favour tillage crops or grass (Figs. 3,4). The heavy and frequent rainfall of Wales, the Lake District and the south-west, most of which have over 40 inches, makes conditions hazardous for harvesting crops, particularly cereals. Heavy rainfall is not even advantageous for grass since it makes for acid soils; as a result the quality of the grass deteriorates and has to be restored by periodic reseeding. Over eastern England rainfall is sufficient for the growth of tillage crops, and the smaller number of rain days improves the chances of a successful harvest, although thunderstorms, which are most frequent in eastern England in summer, often badly affect standing crops. Annual values are misleading, because one of the distinguishing characteristics of the British climate is its variability. Conditions are rarely ideal for harvesting, even in eastern England, and can be a decided handicap in wet summers such as that of 1958. Relatively speaking, however, rainfall conditions in eastern England, where there are fewer than 30 inches per annum, favour the growing of crops other than grass and, conversely, are thought to be too dry in most years for the growth of good grass on light soils.

Heavy rainfall in the west is reinforced by high humidity (which may also be important in the spread of some crop diseases), and by the smaller loss of rainfall through evaporation and transpiration, so that the differences in effective rainfall are even greater than the map suggests. Soils in the north and west rarely dry out sufficiently in summer for water shortage to be a serious hazard. In eastern areas, however, not only is the rainfall lower, but water loss is greater, so that soils generally dry out in summer and plants often lack sufficient moisture for maximum growth, particularly in spring and early summer. A measure of this deficiency is the frequency with which water could be usefully applied to crops. Figure 5 shows the frequency of irrigation need for a green crop; along the east coast such conditions obtain at least 9 years in 10 and, over the whole of East Anglia and south-east England, at least 7 years in 10. The need is less for cereals and sugar-beet and greater for grass, vegetables and potatoes. The areas actually irrigated are, of course, much smaller than those to which water could usefully be applied on theoretical grounds, and were estimated at 266,000 acres in 1965, largely in the vegetable growing areas.

Rainfall does not have any important direct consequences for livestock, except in the provision of drinking water. It is calculated that cattle require some 10 gallons of water per head per day, but much of this is now obtained from public supplies; for it is estimated that two-thirds of agricultural holdings in England and Wales are supplied from public mains.

Temperature conditions affect both crops and livestock. Low temperatures, especially when accompanied by high humidities, make it necessary to house many kinds of livestock. Dairy cattle generally need protection in winter, at least at night, and sample surveys by the Ministry of Agriculture and the Milk Marketing Board have shown that the length of time they are housed varies considerably throughout the country; in southern England the majority of herds are not housed at night until December, while in northern England most herds are housed from late October for 6 months or more. Winter temperatures are also of great importance to hill sheep, determining not only whether they need supplementary food, but even whether they can survive; severe winters in the upland can decimate sheep populations.

For crops, the temperature regime determines both the onset of active growth and also the rate of growth. Thus the beginning of the growing season for grass, approximating to the onset of a mean air temperature of 42°F. (6°C.) and above, occurs before 1 March along the coasts of Wales and the south-west peninsula and a month later in upland Wales and the Pennines. Similarly it ends before November in the uplands and may extend beyond Christmas in the south-west. The growing season is thus longest along the west and south coasts and shortest in the uplands (Fig. 8); in northern England it is estimated to fall by 10 days for every 260 feet of altitude. Figure 8 has been computed on the basis of mean monthly temperatures, and while there may be occasional falls in daily mean temperatures below 42°F. in late spring and early autumn, these are unlikely to damage in any way the growth of an established sward.

Equally as important are the amounts by which temperatures exceed the minimum for growth, for the advantages of an early start may be offset by a slow rate of progress; thus the strawberry crop can begin 2 weeks earlier in the south-west than in the Evesham area, but this lead is soon lost owing to slower rate of growth. A somewhat better measure is given by the total number of degrees by which mean daily temperatures throughout the growing season exceed 42°F. (Fig. 6). Most of southern Britain, excluding the higher uplands, has more than 2,500 day degrees, a figure which falls to less than 2,000 in northern England. For cereals the actual growing season is less than the theoretical, for the crops are harvested in August and September; the margin of advantage enjoyed by eastern districts is thus greater than the map of accumulated temperatures shows.

Many other climatic considerations are relevant in crop growth, although their significance is often local. Sunshine is important for the ripening of fruit and in glasshouse cultivation. Most of Kent and the whole of the south coast receive an average of more than $4\frac{1}{2}$ hours of bright sunshine per day throughout the year (Fig. 7). Sunshine in and near large towns and industrial areas is reduced by smoke and atmospheric pollution. Around the industrial towns on the Pennine flanks pollution

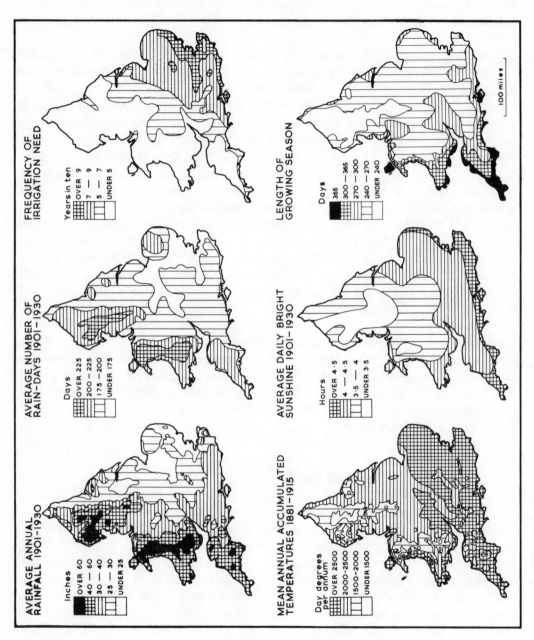

Figs. 3–8. Climate.

deposits exceed 40 tons per square kilometre, making for high soil acidity and poor grass. In the case of the Lea glasshouse industry, pollution has seriously reduced sunshine, while costs of cleaning glass after severe smog exceeded £16 per acre in the 1950s.

Frost is another climatic hazard, of particular importance to the fruit grower and the market gardener. In a broad view, the incidence of frost is least around the coast and increases inland and with altitude; but such information is of little use to the horticulturalist, for topography plays an important part in determining the likelihood of late frosts. The risk is greatest where cold air is likely to accumulate; thus it is less on the Lower Greensand escarpment in East Anglia than in surrounding lowlands, while even in the west, there are low-lying areas like the Plain of Somerset where the risk is high. The grower's aim must be to secure a site with good air drainage where frost risk is at a minimum. Wind also handicaps the farmer. The windiest areas lie around the coasts of the south-west peninsula and South Wales and in the uplands; tree growth is difficult in such areas, farm buildings must be securely roofed and shelter has often to be provided for livestock. Local conditions, especially the roughness of the terrain, greatly affect the speed of the wind, and in flat areas of light soil, particularly the Fenland and the Breckland, soil erosion can be a considerable hazard, involving farmers in considerable expenditure in replanting crops and cleaning out dykes.

It is important to remember not only that the agricultural significance of climate has not been fully investigated and that conditions vary very much from locality to locality, but also that climatic factors do not act independently, while their effects may be reinforced or diminished by conditions of soil and topography. Adequate moisture is of little use without appropriate temperatures for plant growth, while the effect of heavy precipitation may be partly offset by a free-draining soil, as in the Vale of Eden and south-west Lancashire.

Soils

Differences in soil texture, drainage and fertility are of major importance in explaining regional contrasts in agriculture. Light soils warm up quickly and are easy to work at most times, but are low in plant nutrients and droughty. Heavy soils are most costly to work; their cultivation also involves a much greater element of risk than does that of lighter soils, for they can be worked only for limited periods during the year. Heavy soils generally make good grass in drier parts of the country, while grass on light soils is liable to burn; on the other hand, heavy soils poach badly in winter and thereby restrict the grazing season. The significance of many of these soil differences will emerge in subsequent chapters.

The soils of England and Wales have only partly been mapped in detail and, while Figure 9 shows the generalized distribution of the main types, it must be appreciated that each area contains a variety of soils which may differ considerably from each other. The soils shown may be broadly divided into three groups: soils of the uplands, the free-draining soils of the lowlands and the gleyed soils of the lowlands.

Most of the areas above 1,000 feet are mapped mainly as podsols, many of which suffer from impeded drainage. In North Wales and the Pennines there are also areas of peat, much of it eroding, and in the Lake District and Snowdonia smaller areas of thin skeletal soils. Such soils of low base status are of little agricultural value and are mainly under moorland. The lower hill country is largely occupied by acid brown

Fig. 9. Generalized soil types.

Key: 1, soils derived from lowland peat and alluvium; 2, upland peat and podsols; 3, calcareous soils; 4, lowland podsols; 5, acid brown soils; 6, grey-brown podsolic soils; 7, grey-brown podsolic and brown forest soils; 8, gleying occurs in these areas.

forest soils developed on non-calcareous rocks: many of these soils, except in much of the south-west, suffer from impeded drainage, but in some areas, notable the Peak, there are calcareous soils which are free-draining and of greater value to the farmer.

Lowland podsols are restricted in extent and are developed mainly on coarse-grained parent materials such as the Bagshot Beds of the Thames valley, the Hythe Beds of the Weald, the Tertiary sands and gravels of the New Forest and the Triassic sandstones of the west Midlands. Such soils are of little use for agriculture and large areas are under heath and conifer plantations. Rendzinas and other calcareous soils are found mainly on the outcrops of the Magnesian Limestone and of the Jurassic limestones, especially in the Cotswolds, and on the chalklands of the Lincolnshire and Yorkshire Wolds and of the downs in Hampshire, Wiltshire, Dorset and Berkshire. These are generally thin soils, though they are easily ploughed. Other free-draining soils in the lowlands are mapped as brown forest soils or as grey-brown podsolics. The most extensive areas are to be found on the Old Red Sandstone of the Plain of Hereford, the Triassic rocks of the lower Trent valley and the glacial loams and sands of Norfolk. Although they are not all good agricultural soils, they can be ploughed without difficulty and have generally been under arable cultivation. Man-made drainage has converted the peats of the lowlands into highly productive soils, notably the fen peats of the Fenland, but also the mosslands of southern Lancashire. The main areas of alluvial soils, the northern Fenland, the Somerset levels, the warplands of the Humber and Romney Marsh, also owe their high agricultural value to man-made drainage channels and coastal embankments.

Throughout much of the remainder of the lowland, soils suffering from impeded drainage occur; they are either gleyed grey-brown podsolic soils, as over much of the Midlands and northern England, or gleyed grey-brown podsolics and brown forest soils, as in the clay vales flanking the Chalk and Jurassic escarpments. With adequate drainage such soils can be productive arable land, but most are soils of heavy texture and tend to be left in grass.

Over roughly half the lowlands and over nearly all the uplands soils suffering from impeded drainage are common, although interspersed with freely-draining soils. On the other hand, calcareous soils are rare in the uplands and account for less than a fifth of the lowland soils. Free-draining soils of good base status are thus of limited occurrence.

Land Classification

In view of the marked difference in soil and relief all agricultural land is clearly not of equal value. It is difficult, however, to assess these differences because there is no absolute scale which can be used. The ideal soil for grass is not necessarily the best soil for tillage crops, nor are the most suitable soils for wheat and barley identical. Furthermore, since climatic conditions are not the same everywhere, a soil which drains too freely in the east may have advantages in areas of high rainfall in the west. It is also difficult to take account of differences in both the quality and the quantity of produce; thus, although the peats of the Fenland give some of the heaviest yields in the country, they do not produce crops of the highest quality.

Figure 10 presents, in simplified form, the Land Utilisation Survey's map of land classification, the only attempt which has so far been made to produce such a map for the whole country. The classification is based partly on the physical characteristics of the land and partly on the use which it has been put. The map inevitably reflects conditions in the 1930s in England and Wales when much of the light land was

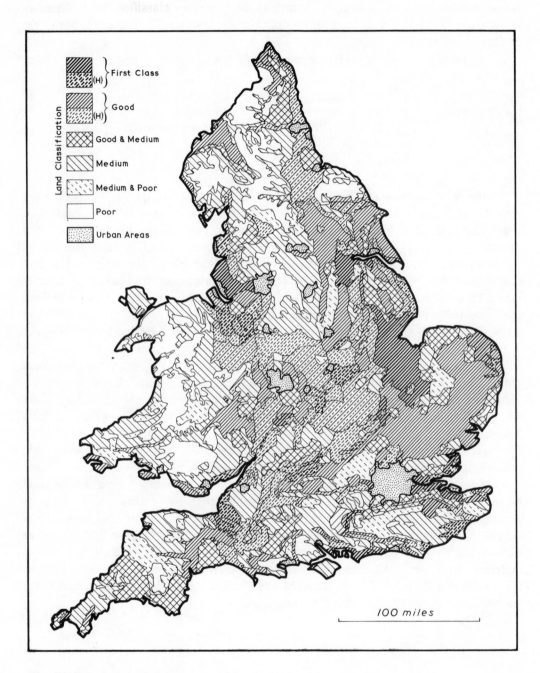

Land Classification

First Class
Good
Good & Medium
Medium
Medium & Poor
Poor
Urban Areas

100 miles

Fig. 10. Land classification.

43

suffering from 50 years of neglect. A classification begun by staff of the Ministry of Agriculture in the 1960s and almost complete gives considerably higher ratings to the chalklands.

L. D. Stamp gives the following proportions of the total area of England and Wales:

	%		%		%
1st Class	5·3	Medium (Light)	7·0	Moorland	12·1
1st Class (Heavy)	3·2	Other Medium	25·0	Other Poor land	4·9
Good	26·1				
Good (Heavy)	13·3				
Good	47·9	Medium	32·0	Poor	17·0

The scheme of shading has been devised to show decreasing quality by decreasing intensity of shading. Good- and first-class land is gently sloping or level land, with deep soils and favourable drainage conditions; it consists mainly of loams, but includes some other textural groups. Both good- and first-class land have been subdivided on the basis of their suitability for grass or arable in 1939, the medium and lighter land then being generally under arable cultivation and the heavier land under grass. This distinction is not wholly satisfactory for there have been changes in land use since 1939; thus a much larger extent of Romney Marsh is now ploughed than in 1939. The chief areas of first-class arable land are the Fenland, the Humber warplands, the loams and peats of south-west Lancashire, and the loams of the Thames terraces, along the dip slope of the North Downs and along the Sussex coast plain; other areas of smaller extent mark the Biggleswade and Evesham market gardening regions. The chief areas of first-class (heavy) land are the Cheshire Plain and the Somerset Levels, Romney Marsh and parts of the clay vales of the East Midlands. Good quality land is extensive in East Anglia, covering most of High Norfolk, Suffolk and Essex, and in the Vale of York, the Lower Trent valley, the Vale of Eden, the Plain of Hereford and the red lands of east Devon.

Medium quality land is land capable of only medium productivity even under good management, and is handicapped by unfavourable relief, by shallow soil or by poor drainage. It thus includes most of the remainder of Lowland Britain and also a large part of the hill margin. Poor land suffers from low productivity, arising from excessive elevation, soils of extremely heavy or light texture, or waterlogging. It is confined largely to land over 1,000 feet and to certain problem areas in the lowlands, such as the 'moorlands' on the Culm Measures of west Devon, the outcrop of the Weald Clay, and the sands of the Breckland and the Dukeries.

Physical Regions

Many of the distinctive localities in England and Wales have been given regional names and Figure 11 has been prepared to show some of the more important of these to facilitate identification of features on the agricultural maps. Some of these regions have fairly clear-cut boundaries, but in others the margins are indistinct, particularly where upland grades into lowland, as on the dip slope of the Cotswolds, and the boundaries shown are inevitably subjective. The areas mapped are:—

Upland Areas

A Cheviot
B Northern Pennines
C Central Pennines
D Lake District
E North York Moors
F Bowland Forest
G Southern Pennines
H York, Derby and Notts Coalfield

I Peak District
J Snowdonia
K Mynydd Hiraethog
L Clwydian Range
M Berwyn Mountains
N Plynlymon
O Radnor Forest
P Mynydd Eppynt
Q Black Mountains

R Brecon Beacons
S South Wales Coalfield
T Exmoor
U Quantocks
V Bodmin Moor
W Dartmoor
X Penwith
Y Wendron Moors

Lowland Areas

1 Tweed Valley
2 Solway Lowlands
3 Tyne Gap
4 Eden Valley
5 Teesdale
6 Durham Coalfield
7 Furness
8 Wensleydale
9 Cleveland
10 Fylde
11 Craven
12 Wharfedale
13 Vale of York
14 Vale of Pickering
15 Yorkshire Wolds
16 Holderness
17 Humber warplands
18 Isle of Axholme
19 Vale of Conway
20 Vale of Clwyd
21 Wirral
22 Cheshire Plain
23 North Staffs Coalfield
24 Sherwood Forest
25 Lincoln Edge
26 Lincolnshire Wolds
27 Lincolnshire Marsh

28 Lleyn
29 Dovey Valley
30 Vale of Powis
31 Shropshire Hills
32 Cannock Chase
33 Vale of Trent
34 Charnwood Forest
35 Silt Fenland
36 Peat Fenland
37 Midland Plateau
38 Breckland
39 Broads
40 Vale of Teifi
41 Upper Usk Valley
42 Plain of Hereford
43 Malverns
44 Vale of Evesham
45 Edge Hill
46 Sandings
47 Vale of Towy
48 Gower
49 Vale of Glamorgan
50 Lower Usk Valley
51 Monmouthshire Levels
52 Forest of Dean
53 Vale of Berkeley
54 Cotswolds

55 Vale of White Horse
56 Vale of Aylesbury
57 Chilterns
58 Vale of St. Albans
59 Somerset Levels
60 Mendips
61 Vale of Pewsey
62 Marlborough Downs
63 Berkshire Downs
64 Salisbury Plain
65 Bagshot Plateau
66 Thames Terraces
67 North Downs
68 Isle of Thanet
69 Blackdown Hills
70 Blackmore Vale
71 Cranborne Chase
72 Dorset Heaths
73 New Forest
74 Surrey Hills
75 High Weald
76 Vale of Kent
77 Vale of Sussex
78 Pevensey Levels
79 Romney Marsh
80 South Downs

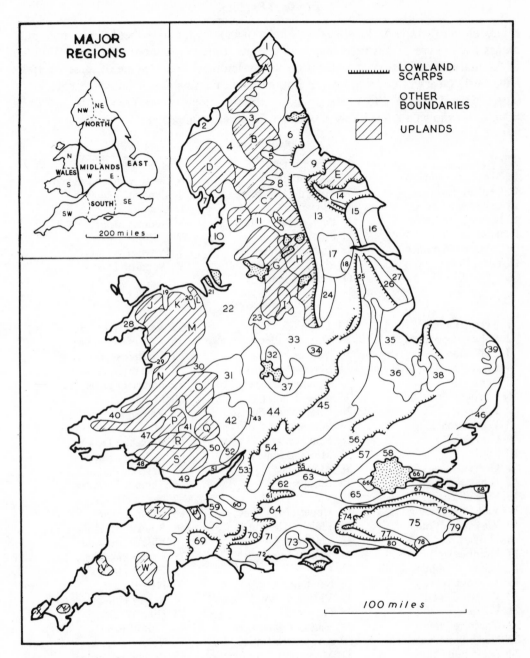

Fig. 11. Physical regions.

Fig. 12. Agricultural Development and Advisory Service Districts 1970.

District Boundaries

Figure 12 records county and district boundaries as they existed in 1970. Comparison with Figure 11 will show that some districts correspond with natural divisions while others cut across physical boundaries. The following districts largely correspond to or are contained within natural divisions:

1	Till-Tweed Valley	76 Wirral	278 Exmoor
13	Solway Lowlands	80A North Staffs Coalfield	286 Blackdown Hills
25	Teesdale	123 Sherwood Forest	289 Dartmoor
32	Holderness	136 Vale of Ancholme	299 Land's End Peninsula
37	Cleveland	165 Broads	303 Lleyn Peninsula
38	North York Moors	237 Romney Marsh	327 Vale of Teifi
44	Craven Lowlands	253 New Forest	342 Vale of Glamorgan

The following groups of districts correspond roughly with certain major natural divisions, though they include some contrasting terrain:

4, 5	Cheviot foothills	138, 144, 145, 147, 148	Jurassic Escarpment
26, 27	Yorkshire Wolds	161, 163	Norfolk Loam Region
35A, 36, 41, 49	Vale of York	159, 160, 178	Breckland
43, 45, 46	Central Pennines	201, 213	Chilterns
52, 61	Southern Pennines	239, 240	High Weald
29, 56, 59	Warplands	247, 248, 249, 267	Chalk Uplands of Hampshire
53, 62, 65, 117, 117A	Yorkshire Coalfield		and Wiltshire
82A, 119, 122, 124	Trent Valley	256, 257	Cotswolds
91, 91A	Shropshire Hills	279, 279A	Somerset Levels
94, 95, 96	Plain of Hereford	281, 282	Moorlands of West Devon
97, 98, 100, 102, 105	Vale of Evesham	304, 305, 306, 307	Snowdonia
111, 112, 113	Upper Avon Valley	308, 315, 316, 317	North Wales Uplands
135, 153, 154, 155, 156, 166, 189	Silt Fenland		

The districts in each county are: Bedford, 196–8; Berkshire, 221–5; Buckingham, 211–15; Cambridge and Isle of Ely, 184–91; Cheshire, 76–9; Cornwall, 291–9; Cumberland, 12–17; Derby, 116–20; Devon, 280–90; Dorset, 268–72; Durham, 20–5; Essex, 202–10; Gloucester, 255–61; Hampshire, 247–53; Isle of Wight, 254; Hereford, 92–100; Hertford, 199–201; Huntingdon and Soke of Peterborough, 192–5; Kent, 232–7; Lancashire, 67–71; Leicester, 139–43; Lincoln (Holland), 153–7; Lincoln (Kesteven), 125–8; Lincoln (Lindsey), 129–36; London, 226; Norfolk, 158–70; Northampton, 144–8; Northumberland, 1–7; Nottingham, 121–4; Oxford, 216–20; Rutland, 138; Shropshire, 86–91; Somerset, 273–9; Stafford, 80–5; Suffolk, 173–83; Surrey, 228–30; Sussex (East), 238–42; Sussex (West), 244–6; Warwick, 109–15; Westmorland, 18–19; Wiltshire, 262–7; Worcester, 101–6; Yorkshire (East Riding), 26–33; Yorkshire (North Riding), 34–41; Yorkshire (West Riding), 43–65; Anglesey, 300–2; Brecon, 328–30; Caernarvon, 303–7; Cardigan, 324–7; Carmarthen, 344–7; Denbigh, 308–11; Flint, 312–14; Glamorgan, 338–43; Merioneth, 315–16; Monmouth, 332–6; Montgomery, 317–20; Pembroke, 348–52; Radnor, 321–3. Not all these districts are shown on Figure 12, as some have disappeared through reorganization.

CHAPTER III

The Man-made Framework of Farming

The land is not farmed in a state of nature. It is divided into fields and farms, equipped with farm buildings and machinery, and farmed by a large number of men of varying ability and outlook. The size, shape and layout of farms and fields, the kind of equipment, the availability of labour, the location of markets and the attitudes of farmers all influence the way in which the agricultural land is used. This chapter examines some of the available data relating to these questions, although it must be admitted that much of the information is uneven in quality and available only for counties or larger units.

British farms form a complex mosaic of interlocking pieces of varying sizes, but there is little satisfactory contemporary information about their shapes or sizes. While most farms lie in one block surrounded by a ring fence, in some parts of the country farms are often fragmented, notably in the Midlands where many farms were laid out in the period of Parliamentary enclosure between 1750 and 1850, and in areas of small holdings, like the Fens and the Evesham district. The National Farm Survey of 1941–3, the only source of data on farm layout for any large area, showed that while 75% of the holdings in 24 counties were in one piece, the proportion of severed holdings ranged from under 5% in Devon and Cornwall to 50% and over in Cambridge and Nottingham. In some areas the layout of farms is related to relief and soils; for example, in parts of the scarplands farms are often laid out at right angles to the grain of the country and so include a variety of terrain. Similar layouts are frequently found in hill country, where farms include both in-bye land in the valley bottom and rough grazing on the open moorland above the valley. In addition, upland and lowland farms, or farms on contrasting terrain, are often linked by the same management and so complement each other, the upland farm, for example, providing summer grazing while the lowland farm grows feed crops and accommodates stock in winter. There are insufficient data to map the regional distribution of such differences in layout, but they are clearly important for the interpretation of these district maps.

There are also regional differences in field size which exercise some influence over the use of land, but, while the size of field is given on the Ordnance Survey 25" Plans, these vary considerably in date of survey, and no information exists from which maps of the country can be drawn showing the present position. In eastern arable districts, particularly on the outcrop of the Chalk, fields are often large and hedgeless, conditions which facilitate mechanized corn growing; in one sample in south

Oxfordshire over 74% of the land was in fields of 30 acres and over. In the west and north, fields are generally small and separated by massive banks or stone walls; in a survey of farms on Exmoor it was found that half the land on farms of 200 acres and over was in fields of under 7 acres, while on farms under 50 acres the proportion was as high as five-sixths. Such small fields are quite well suited to livestock farming, providing shelter and facilitating controlled grazing, but they are a considerable handicap in mechanized arable farming and result in higher costs. Sample data collected by the Forestry Commission on lengths of field boundaries give an indirect measure of regional differences in field size. In Devon there were over 20 miles of boundaries per square mile, while in Essex there were a little over 12, suggesting that fields in Devon are on average about one-quarter the size of those in Essex; the existence of areas of open moorland in Devon certainly minimizes the difference between the two counties.

There are similar differences throughout the country in the character of farm buildings, a reflection of local materials, climatic conditions and type of farming. Buildings represent a large proportion of the fixed capital on farms, and farming is increasingly going under cover to provide shelter for livestock, harvested crops and implements. Unfortunately data on regional variations in farm buildings and their significance are lacking.

Most of the mappable information about the man-made framework is derived incidentally from the agricultural returns. It is important to notice that these returns are obtained, not from farms, but from all holdings of more than 1 acre of agricultural land. Many of the smaller holdings are not farms in the accepted sense, but accommodation fields, grounds of large institutions and similar parcels of land used for agricultural and semi-agricultural purposes. Many of these holdings were statistically insignificant for agricultural purposes, i.e., they were under 10 acres of crops and grass, employed no regular whole-time labour and required fewer than 26 standard man-days, and some 47,000 such holdings were excluded from the census in 1968. Some holdings are in effect parts of a larger unit, but have continued to be returned separately either through inertia or because the occupier wished to do so. In 1970 about 10,000 such holdings were amalgamated with their parent holdings for statistical purposes. At the same time, some 2,000 holdings with a significant agricultural output but with an area of 1 acre or less have been included in the census (though there may well be other such holdings not known to the Ministry of Agriculture's staff). Nevertheless, despite these reforms, there are still many holdings which were occupied on a part-time or spare-time basis and consequently many more holdings than there are farmers returned in the population census (though the contribution of these small holdings is estimated at only about 7% of agricultural output).

Ownership and Occupation of Agricultural Land

Tenant farming has long been thought to be the characteristic feature of the occupation of agricultural land in England and Wales, but this view now requires considerable modification. Since the First World War there has been a marked increase in the percentage of land owned by occupiers. Over the whole of England and Wales, in 1970, some 53% of agricultural land in sole ownership (i.e., crops, grass and sole ownership rough grazings, but excluding common land) was owned by those who occupied it. Proportions of owner-occupied land were highest in Wales and southern England, and lowest in the north and Midlands (Fig. 13); over 60% of farmland in

Fig. 13. Land occupied by owners as a percentage of agricultural land (excluding common rough grazing).

Fig. 14. Part-time holdings as a percentage of all holdings.

Wales was owner-occupied (Radnor 68%), whereas more than half the farmland in north England was tenanted (Northumberland 62%). In 1950 only 38% of farmland was owner-occupied and in Westmorland and Durham the proportion was less than a quarter; nowhere was it higher than 56%. This growth of owner-occupation is due in part to the break-up of large agricultural estates and the sale of constituent farms to sitting tenants, to difficulties in obtaining a tenancy and to a growing interest in part-time and hobby farming, though interpretation of the data is complicated by the conversion of some large estates into private companies, by the growing popularity of lease-back arrangements, whereby owner-occupiers become tenants again, and by the acquisition of estates by large institutional investors.

There are thus considerable differences in the degree to which occupiers of agricultural land depend on their holdings for a livelihood. The first complete source of evidence on this topic also comes from the National Farm Survey. This showed that while in England and Wales as a whole 94% of the land under crops and grass was occupied by those who depended wholly or partly on their farms for a living, proportions ranged from 99% in Holland and the Isle of Ely to 78% in Surrey, and were lowest in the East Midlands and the Home Counties.

More recently, J. Ashton, B. E. Cracknell and others have examined a related question, using data from the June censuses and standard labour requirements to estimate the number of holdings which did not provide their occupiers with full-time employment. Since they were concerned with the classification of holdings they did not distinguish spare-time, part-time and hobby farmers, all holdings providing less than 275 man-days of employment per year being regarded as part-time (although some 10,500 of these proved to be the sole source of livelihood of their occupiers). Since 48,000 of the 180,000 part-time holdings provided less than 25 man-days of work, while occupiers of nearly a third of the remainder had other full-time occupations, it is clear that many of these holdings would have been occupied by spare-time farmers in the National Farm Survey sense, i.e., those with other full-time employment. There are now estimated to be 104,000 part-time holdings, or 45% of the total, though this figure is not comparable with previous estimates for the reasons already outlined (p. 50).

The distribution of these part-time holdings presents both expectable and surprising features (Fig. 14). High proportions are associated with large urban areas and with districts where intensive horticulture is practised; they are also found in many parts of Wales, where they are probably both a legacy of the past, when part-time farming was often associated with mining, and a reflection of the present unfavourable structure of farming in many areas. By contrast, proportions are much lower in northern England. These differences should be interpreted with caution, as they relate only to proportions of holdings, but they are nevertheless confirmed by computation of the proportion of the agricultural area occupied by such holdings.

Ashton and Cracknell also examined the legal status of occupiers of all holdings. Not surprisingly, in view of the numerical superiority of small holdings, they found that 88% of all holdings were occupied by individuals and a further 9% by partnerships, many of which would probably be between brothers, or between father and son. Joint stock companies (2%) and other occupiers (Government departments, public bodies and the like) account for only a small proportion of total holdings, but they are probably more common in the larger-size groups. Proportions of occupiers other than partnerships and individuals are highest in the eastern and south-eastern counties and lowest in Wales.

The effect of these differences in the status of occupiers are difficult to evaluate. Part-time holdings, as indeed small holdings in general, are more likely to concentrate on certain enterprises. It also seems probable that where hobby and spare-time farmers are more numerous, farming will be more varied than it would otherwise be; for such occupiers are both freer to indulge their personal whims, and more likely to do so since they are probably less hidebound by traditional practices. Indeed, hobby farmers are often the innovators in farming.

Farm Size

Another important consideration is regional variation in farm size. Farm size has been shown to be related to the kind of enterprise; in particular, farming on small farms is likely to be more intensive than that on large, and to be concentrated on the production of milk, vegetables, pigs and poultry, which demand considerable inputs of labour, purchased feed or capital and which are therefore more likely to provide an adequate income on a small acreage; 64% of the egg-laying flock, 51% of breeding fowl and 36% of the dairy herd were on holdings of less than 100 acres of crops and grass in 1970, although these holdings accounted for only 22% of the acreage under crops and grass; By contrast, such holdings had only 6% of the acreage under wheat. Extensive enterprises, such as sheep rearing, cannot provide their occupiers with a satisfactory income from only a small acreage; thus, while $\frac{1}{2}$ an acre of grass or 10 acres of top fruit can support a family, farms, other than specialists, with less than 50 acres generally show a net loss when allowance is made for the farmer's labour and capital. Although there are few economies of scale in livestock rearing, holdings must necessarily be fairly large because of the low returns per acre, unless occupiers are prepared to accept a low standard of living. In cereal growing, on the other hand, which can be highly mechanized, farms tend to be large because there are economies of scale.

The size of farms may be measured in a number of ways. Historically, the only measure available has been the area in crops and grass, but it is now possible to consider both the total acreage of agricultural land and agricultural output, as measured by standard labour requirements. Each approach has advantages and disadvantages. Total acreage gives the most useful measure of areal extent (though the varying amount of common rough grazing, not included in estimates of farm size, should be kept in mind; see Fig. 47); on the other hand, it places an equal value on all farmland. The use of acreages of crops and grass and labour requirements avoids this difficulty, but may give a misleading visual impression, especially in respect of intensive farms, such as specialist poultry and horticultural holdings. Figures 15 and 16 show these contrasting approaches by recording the average size of holding in each district, as measured by total acreage and total man-days respectively. On Figure 15, high values appear in two contrasting areas, the uplands, especially those of northern England, and the chalklands and other limestone escarpments of lowland England; the lowest values are associated mainly with the major cities and with areas of intensive horticulture. The map of the average size by man-days also shows the chalklands as the principal area with large holdings, but the uplands are now characterized by the smallest holding, an indication of the low intensity of the farming systems practised there (Fig. 16).

53

A more useful indicator of the distribution of farms of different sizes is the proportion of farmland occupied by each size group. For present purposes, six size groups, as measured by area of agricultural land, were identified, viz., under 20 acres; 20–99 acres; 100–299 acres; 300–499 acres; 500–999 acres; and 1,000 acres and over. Such holdings occupy 2%, 17%, 36%, 17%, 16% and 12% respectively of the agricultural area. Owing to the very different proportions of the agricultural area occupied by holdings in the various size groups, it has not been possible to use the same class intervals. Great care must therefore be taken in comparing these maps, for

Fig. 15. Mean size of holdings (acreage of agricultural land).

Fig. 16. Mean size of holdings (standard man days).

while each shows those areas in which that particular size group is relatively important or unimportant, the actual percentages represented by the various shadings are very different; thus, whereas the top category on Figure 17 represents 5% and over, the same shading represents 40% on Figure 18 and 60% on Figure 24.

Small holdings of less than 20 acres are, as might be expected from the preceding analysis of average size, especially prominent around the cities and in the main horticultural areas (Fig. 17). Holdings of 20–99 acres occur in three principal areas: those where small-scale dairy and poultry enterprises are important; south-west Wales, an area of mainly dairying; and west Cornwall, where horticultural crops are important (Fig. 18). The lowest values occur in the uplands and in the main arable areas. Holdings of 100–299 acres, which occupy a third of all farmland, are widespread, with no area pre-eminent; the uplands and the main arable areas have the lowest proportions of holdings in this category (Fig. 19). With successively larger holdings, the pre-eminence of the main arable areas and (to a lesser degree) the uplands becomes steadily more marked (Figs. 20–22). Holdings of 1,000 acres and over occupy half or more of the agricultural land over large tracts of north Northumberland and the chalklands of west Norfolk and Hampshire (Fig. 22). These maps also demonstrate the difficulty of interpreting such complex data when a very large number of spatial units is employed.

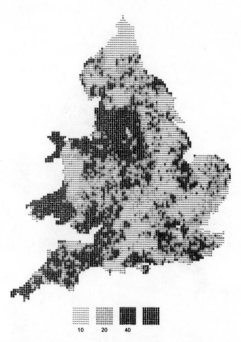

Fig. 17. Holdings under 20 acres as a percentage of agricultural land (excluding common rough grazing).

Fig. 18. Holdings of 20–99 acres as a percentage of agricultural land (excluding common rough grazing).

Because historical data about farm size in England and Wales relate only to measurements by acreage under crops and grass, interpretation of change is difficult. Nevertheless, it seems fairly clear that both the number and acreage of larger holdings have been increasing; between 1944 and 1970 the number and the acreage of holdings of 500 acres and over more than doubled, while those of 1,000 acres and over more than trebled. The regional pattern on the other hand, seems to have changed little.

Rent of Agricultural Land

Although less than half the agricultural land is rented, rent is of particular importance to the tenant farmer because it represents an overhead which has to be met. In 1970 rents accounted for 9% of all farm expenses, or 14% after the exclusion of feeding stuffs, livestock and seeds, a value that varies between 7% on horticultural holdings and 22% on cropping (mostly cereals) farms in the Farm Management Survey sample. Rents have been rising rapidly in the period since 1958 and have certainly been a factor in the intensification of farming systems.

Unfortunately there are no statutory returns of rent, and in any case some 46% of holdings are owner-occupied. Full data from all holdings over 5 acres were collected

Fig. 19. Holdings of 100–299 acres as a percentage of agricultural land (excluding common rough grazing).

Fig. 20. Holdings of 300–499 acres as a percentage of agricultural land (excluding common rough grazing).

in the National Farm Survey 1941–3 and subsequent sample surveys have been made by the Agricultural Land Service. Data from the 1970 survey were obtained from this source for the construction of Figure 23, which shows average rent levels by counties.

Agricultural rents may be regarded as an integration of all the preceding factors, for rents vary with the quality of land, with size of holding and with location. Rent per acre tends to fall with increasing size of holding, for the value of the residence and buildings bulks much larger on small holdings; the Land Service's survey found that the average rent in 1970 fell from £9.00 on holdings of less than 25 acres, to £5.96 on holdings between 50 and 100 acres and £4.08 on holdings of 1,000 acres and over, compared with an average of £6.11. Areas where small holdings predominate are thus likely to have higher rents than are those where large holdings are numerous. Rents may also be high around towns, reflecting advantages of location, possibilities of urban expansion and the existence of numerous, small, semi-agricultural holdings. Rents also vary with the quality of land and with type of farming, though these are closely related. In the 1970 Farm Management Survey average rentals per acre ranged from £1.10 on livestock, mostly sheep farms, to £7.00 on general cropping farms and £12.10 on horticultural farms.

Fig. 21. Holdings of 500–999 acres as a percentage of agricultural land (excluding common rough grazing).

Fig. 22. Holdings of 1,000 acres and over as a percentage of agricultural land (excluding common rough grazing).

The distribution of average rents per acre of crops, grass and rough grazing is shown on a county basis in Figure 23, but it must be remembered that there are considerable differences within counties; in Buckinghamshire, for example, rents in 1941 were almost twice as high in the clay vales as in the Chilterns. The county pattern of rents is primarily a reflection of land quality and, to a lesser extent, of type of farming; for, in general, where average rents are high or low they are high or low for all types of farming. Rents over £7.00 per acre are found in two contrasting areas. In Cheshire much of the land is good quality grassland, farmed intensively in fairly small dairy farms; in the Fenland, where average rents in 1970 were £7.00 per acre, some of the best arable land in the country is also farmed mainly in small units. Rents between

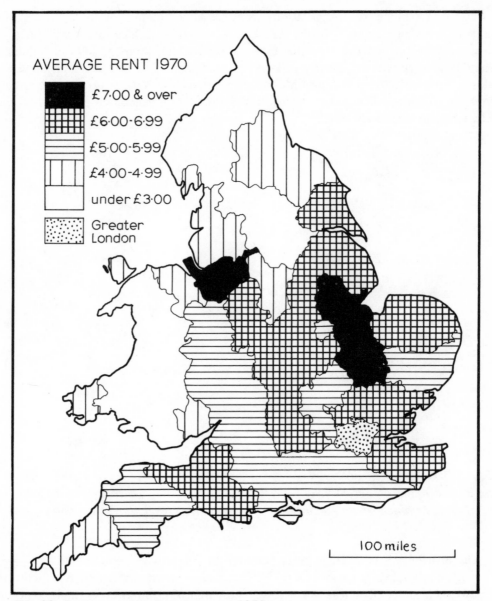

Fig. 23. Average county rents per acre 1970.

£6.00 and £7.00 per acre occur in most counties of eastern England and the Midlands, reflecting the general tendency for rents on arable farms to overtake those of grass farms over the past fifteen years, a reversal of historical trends. Rents are generally lowest in upland counties of Wales and the north of England, in some of which rents are below £3.00 per acre. In these counties there are large areas of rough grazing, while much of the enclosed land consists of poor permanent pasture.

The Agricultural Population

For this purpose it is necessary to refer to the censuses of population, which contain information on the numbers in all forms of employment. Unfortunately, recent censuses do not provide mappable data since data on occupations are collected from a sample of the population, and it is therefore necessary to refer to the map derived from the 1951 census and included as Figure 26 in the first edition of this atlas. Although the agricultural population has been falling steadily, there is no reason to believe that the pattern of its relative importance has changed substantially since.

According to the 1966 census, the agricultural population of England and Wales numbered some 588,620 or 3% of the occupied population, compared with 930,902 or 5% in 1951. Most of those employed in other occupations are, of course, to be found in the towns, for 80% of the population is urban, i.e., lives in urban districts; but even in the rural districts the proportion in agriculture is generally under 50%. Some of the remainder are those who provide services for the agricultural population and hence can be considered part of the rural population proper; S. W. E. Vince has calculated that in 1931 the ratio of agricultural to service employment in strictly rural areas was about 2 to 1. The balance of the population comprises two elements. Many rural districts contain areas which are urban in character, but which, owing to the lag in the adjustment of administrative boundaries, are still classed as rural. Rural areas also contain 'adventitious' population, those who live in the countryside from choice and work in towns; these are most numerous in areas of attractive scenery near large cities, especially London. The agricultural population is thus relatively most numerous in the areas remote from large urban agglomerations and least in the axial belt of industry and population which stretches from Greater London to south Lancashire. Only in upland Wales, the Fenland and parts of the north and south-west England were more than 50% of the employed population engaged in agriculture in 1951, while over much of the Midlands and south-east England and around the industrial areas of South Wales and the north-east coast the proportion fell to under 20%.

When the density of the agricultural population is considered, a somewhat different picture emerges (Fig. 24), although the concept of density must be treated with caution in respect of a labour force whose commitment to agriculture ranges from 100% to under 10%. This map clearly shows that the areas with a predominantly agricultural population are of two contrasting types, the one with a high density of agricultural population, the other with a low density. Thus, while in the Fenland there are more than 50 persons engaged in agriculture and horticulture per 1,000 acres of agricultural land, in upland Wales and northern England densities fall below 15. Conversely, some areas where the proportion of the population employed in agriculture is low have high densities of agricultural population, as in Surrey and Essex. In general, densities are highest in areas of intensive farming, especially in the horticultural areas such as Kent, and to a lesser extent, the dairying areas such as Cheshire, and lowest in areas of livestock rearing, particularly where there is much rough grazing, although the Hampshire-Wiltshire chalklands also stand out.

The composition of the agricultural population itself varies throughout the country (Fig. 26). The proportion of full-time farmers in the full-time agricultural population must clearly be related to farm size and to the type of farming. Where farms are small or where extensive livestock farming is practised the proportion of farmers is high. Over virtually the whole of the uplands farmers, managers and foremen are more numerous than farm-workers, the highest proportion being 70% in District 347B. Over much of eastern and southern England there are at least two full-time farm-workers for every full-time farmer, and the proportion of farmers is generally lowest where farms are largest, although mechanization of arable farming has helped to diminish this contrast between east and west.

When the distribution of full-time farmers is considered, a somewhat different picture emerges (Fig. 25). Full-time farmers are most numerous in areas of intensive farming, where farms are small, especially in areas of specialist horticulture, such as the Fenland and Worcestershire, and fewest where farms are relatively large, as on the chalk and limestone areas of southern England, and in the uplands, although here the labour force is also small.

A further aspect of the composition of the agricultural population is the use of family labour. The agricultural census specifically excludes the contribution of the farmer's wife, but details of family workers, whether full- or part-time, are now obtained. On average, some 31% of the employed regular labour force comprises family workers but there are again marked regional differences. Not only are there relatively few farm-workers in the uplands, but a large proportion of these (over 50%) are family workers (Fig. 27). By contrast, they comprise less than 20% of the regular labour force over most of eastern and southern England. Contrasts would probably be even greater if the contribution of wives was included.

Agricultural Workers

In June 1970, 351,945 workers (excluding the occupiers of holdings and their wives) were employed on agricultural holdings in England and Wales; 74% of these were males. Sixty-two per cent (of whom 90% were males) were employed whole-time, 18% (55% males) part-time, and 20% (50% males) were seasonal and temporary workers. Many of the women are engaged in horticulture, particularly in the Fenland and Kent; regular part-time work in tending and harvesting a succession of crops plays an important part in the economic and social life of the Fenland.

Type of farming, farm size and the ratio of farmers to farm-workers are primarily responsible for regional differences in the number of agricultural workers (Fig. 28). Districts with 26 or more regular full-time workers per 1,000 acres of agricultural land include all the main horticultural areas; the highest density is found in District 228A, with 52 workers per 1,000 acres. The principal regions for dairying and arable farming

Fig. 24. Farm labour per 1,000 acres of agricultural land (excluding common rough grazing).

Fig. 25. Whole-time farmers per 1,000 acres of agricultural land (excluding common rough grazing).

Fig. 26. Whole-time farmers per 100 whole time farmers and workers.))

Fig. 27. Family workers per 100 regular workers.

have 16 or more workers per 1,000 acres; by contrast, there is a belt of districts with below average densities between Wiltshire and Lindsey. Over most of these uplands there are fewer than 10 full-time workers per 1,000 acres, a reflection of their dependence on systems of extensive livestock rearing and of the predominance of small farms which rely mainly on the labour of the farmer and his wife.

June is not a satisfactory time at which to record numbers of temporary workers, which probably reach a peak in the autumn when casual labour for fruit picking and root harvesting is at a premium; there were, for example, some 12,000 more temporary

Fig. 28. Regular workers per 1,000 acres of agricultural land (excluding common rough grazing).

Fig. 29. Seasonal workers per 1,000 acres of agricultural land (excluding common rough grazing).

or seasonal workers in September 1970 than in June. Figure 29 records the distribution of casual workers in June, and shows that numbers are greatest in the areas of intensive horticulture, the highest density being found in District 233 (17 per 1,000 acres of agricultural land).

Numbers of agricultural workers have fallen considerably since the 1930s, although close analysis is made difficult by alterations in the questions asked in the June census. Owing to an increase in the casual labour force, total numbers rose during the Second World War, although the number of regular workers changed little. Throughout the 1950s and 1960s the regular labour force had fallen steadily and was considerably smaller in 1970 than in 1950.

Agricultural Machinery

No change on British farms since the outbreak of the Second World War has been more marked than the vast increase in mechanization. In 1938 there were an estimated 50,000 tractors in England and Wales; by 1970, according to the machinery census, there were some 413,180, of which 61,400 were tractors under 10 h.p. and 7,480 track-laying vehicles. Parallel increases have occurred in other farm machinery. There were only some 50 combine harvesters in 1938, but by 1970 this number had increased to 56,670. The mechanization of sugar-beet harvesting has increased strikingly; in 1946 there were 183 complete sugar-beet harvesters, by 1970 14,920. Numbers of milking machines have also risen rapidly, while there were some 29,370 grain driers of various kinds in 1970, compared with 1,002 in 1946. Since the mid-1950s the number of tractors has ceased to grow and there were in fact fewer in 1970 than in 1956, partly because of the development of more powerful and efficient machines and partly because saturation point had been reached. On the other hand, numbers of more sophisticated equipment continue to rise; those of combine harvesters almost doubled between 1960 and 1970.

Figures 30 to 41 contain some of the many maps which can be drawn to show the regional importance of different types of machinery. They are based on county estimates made from one-third of all holdings, a different third being sampled at each of the three quarterly censuses. While such estimates are generally adequate for national totals and for the preparation of maps of ratios and densities, they are liable to sampling errors, which may be quite large in the smaller counties.

Tractors are most numerous in eastern counties, particularly in areas of intensive farming where there are two or more per 100 acres of crops, grass and rough grazing (Fig. 30); but in nearly all counties there is more than one tractor per 100 acres, and, if the comparison is restricted to the density per 100 acres of crops and grass only, the rest of Wales and north England also have at least one tractor per 100 acres. The percentage of small tractors, which is 15% for England and Wales, varies widely, from 31% in Worcestershire to 4% in Kesteven, the proportion being generally highest in counties with a considerable acreage of horticultural crops (Fig. 31). Large tractors are most strongly associated with areas of large-scale arable farming (Fig. 32). Combine harvesters are also most common in East Anglia, where there are 3 or more per 1,000 acres (Fig. 33); most of these are large machines of 10 ft. cut or wider, but in western counties, with small fields and small acreages under cereals, small machines are relatively more common, as are tractor-drawn combines (Fig. 34). Mowers are likewise most common in western counties, where farms are comparatively small and there are large acreages of grass (Fig. 35). One indicator of change is that binders and horse-drawn cultivators, an index of east-west contrasts in the first edition of this atlas, are no longer enumerated, and the collection of numbers of horses was discontinued in 1959, when there were 72,000, a figure which had fallen to an estimated 19,000 by 1965.

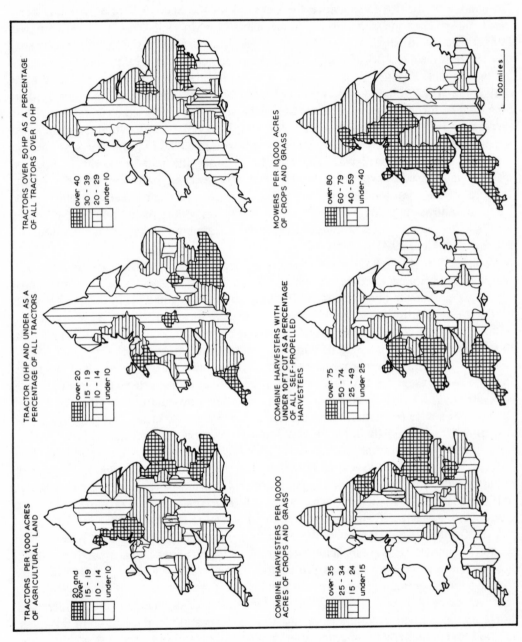

Figs. 30–35. Agricultural machinery.

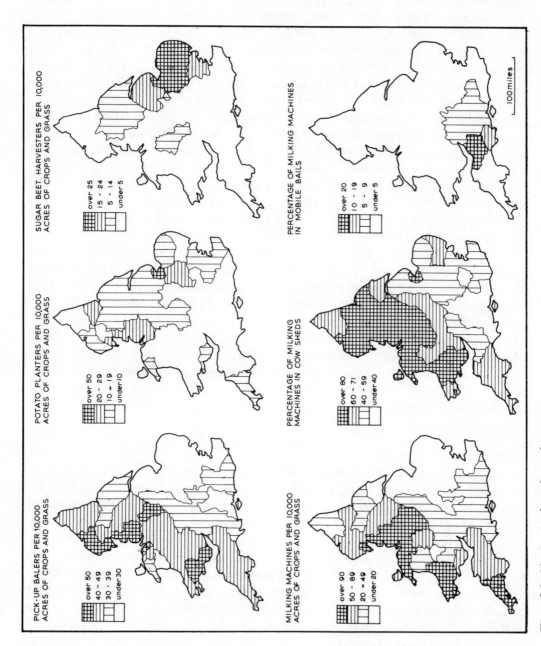

Figs. 36–41. Agricultural machinery.

SUGAR BEET HARVESTERS PER 10,000
ACRES OF CROPS AND GRASS

over 25
15 – 24
5 – 14
under 5

PERCENTAGE OF MILKING MACHINES
IN MOBILE BAILS

over 20
10 – 19
5 – 9
under 5

100 miles

POTATO PLANTERS PER 10,000
ACRES OF CROPS AND GRASS

over 50
20 – 29
10 – 19
under 10

PERCENTAGE OF MILKING
MACHINES IN COW SHEDS

over 80
60 – 71
40 – 59
under 40

PICK-UP BALERS PER 10,000
ACRES OF CROPS AND GRASS

over 50
40 – 49
30 – 39
under 30

MILKING MACHINES PER 10,000
ACRES OF CROPS AND GRASS

over 90
50 – 89
20 – 49
under 20

The remaining maps of farm machinery show other, generally more specialized equipment (Figs. 35–38). In interpreting them it must be remembered that some counties, particularly the larger ones, contain many different types of farm. Pick-up balers are also more numerous in western counties, but potato planters (and also potato-harvesting equipment) are understandably most common in the principal areas for potato production, with the highest density in Holland (Fig. 37); the Eastern Region alone has a quarter of all the machines. Sugar-beet harvesters are even more highly localized, with 40% in the Eastern Region and 20% in Norfolk alone.

Figures 39–41 show aspects of the mechanization of dairying. Comparison of actual numbers of milking machines is handicapped by the wide variety of types; the highest density (treating all types of machine as equal) is in Cheshire, the principal county for dairying. Cowshed milking, like the limited amount of hand milking remaining, is primarily a feature of western counties, although climatic conditions also play a part. This fact is much more evident in Figure 41, which shows the proportion of milking machines which are in mobile bails; this is very much a feature of southern England, with 35% in Somerset alone and 62% in the South-west Region.

Many other aspects of farm machinery could be mapped and discussed, and even for those which have been included, a mirror image of the maps of densities would have been shown if numbers of machines had been computed, not in relation to crops and grass, but in relation to the enterprise for which the equipment is kept, e.g., combines per 1,000 acres of cereals, or milking machines per 100 dairy cows; for ratios are often highest in those areas where acreages or numbers are small, e.g., of mowers to grassland in East Anglia or of combines to acres of cereals in Wales. The choice of different areal bases can exaggerate or minimize east-west contrasts. It is, however, a matter of regret that the data collected about farm machinery cannot be related to other information enumerated in the complete census each June.

Agricultural Markets

In a broad sense, the existence of a large urban population with a rising standard of living determines the character of farming in this country, particularly the emphasis on livestock and horticultural products. But the influence of markets on the location of agricultural production is difficult to evaluate. It might be expected that a bulky crop such as potatoes would be grown around the large urban centres, but this is true only to a limited extent. Pricing policies and the structure of wholesale marketing both distort the effects of location. A few agricultural commodities are sold direct to consumers, but most are sold either to wholesalers or to manufacturers and processors. In some instances the location of factories may be determined by the location of the existing centres of production; in others, the erection of a factory or processing plant may itself give rise to new centres of production. There is, unfortunately, no convenient source from which maps of agricultural markets and factories can be compiled and only a few sample maps can be given.

The siting of sugar-beet factories played an important part in the location of sugar-beet growing areas, though quota limitations on the acreage which may be grown have tended to fossilize the pattern of production; without these there is little doubt that the acreage would be extended. There are 17 sugar factories in England and Wales which process sugar-beet, all but two of them located in the eastern arable areas (Fig. 43). Sugar-beet growers bear the whole cost of transport to the contract factory (except for rail traffic over 40 miles) and the great bulk of sugar-beet is grown within quite a short distance of factories. While the British Sugar Corporation will pay the excess rail freight on beet travelling more than 40 miles, little beet is, in fact, grown beyond this radius of factories, the chief exception being West Sussex. In all, under 1% of the acreage grown in 1970 was located in southern regions, i.e., south of the Thames–Severn line. In practice, movement is not as simple as suggested, for beet may be sent, at the Corporation's expense, to more distant factories, if this ensures more even working at the factories.

The marketing of milk is more complex (Fig. 42); for milk is primarily sold for liquid consumption in the major urban centres. Individual producers do not pay the true costs of transport to the consuming markets so that location in relation to urban markets is not now a major factor in determining the location of milk production. But a variable proportion of the milk produced annually is manufactured into butter, cheese, condensed milk and cream; in 1970–1, 796 million gallons, or 35% of the total, were processed in this way. Manufacture serves as a safety valve for the liquid milk market, taking the surplus which cannot be sold. The proportion of milk going to factories is generally highest in western and northern areas, where there is a tendency to concentrate on the production of cheaper summer milk on grass (see Fig. 171); it is, in fact, Milk Marketing Board policy to rationalize the distribution of manufacture so that milk used in this way does not have to travel long distances and numbers of factories located near urban areas have been closed in recent years. Processing factories are now located mainly in the principal dairying areas where the largest surpluses occur.

Eggs, like milk, are bulky and perishable and are mainly consumed fresh in the principal urban areas. About two-fifths of eggs are sold at the farm gate, chiefly from farms near the major cities, but most of the remainder are marketed through packing stations, of which there were 267 in England and 25 in Wales in 1967, when the Re-organization Commission for Eggs (from whose report this map, Figure 44, was adapted), was investigating. These stations are widely scattered throughout the lowlands, and the policy followed by the Egg Marketing Board, before it was wound up, encouraged this tendency; for packers, who collect eggs from producers, were paid a standard payment for collection and delivery within a 30-mile radius of the packing station, costs of any longer distance to market being met by the Board. The location of these markets was thus of little importance.

The processing of horticultural crops has become increasingly important. Numerous canneries were established in the main fruit and vegetable growing areas in

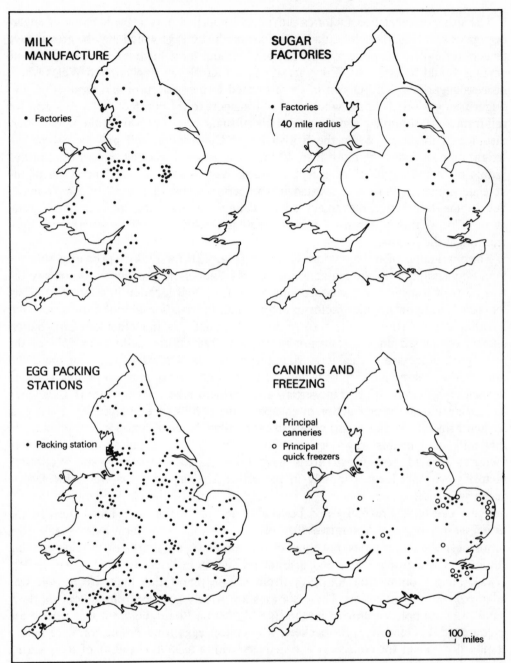

Figs. 42–45. Sample maps of markets and factories.

Fig. 43: Sugar factories are located at Poppleton. Selby, Brigg, Bardney, Kelham, Colwick, Spalding, King's Lynn, Peterborough, Wissington, Cantley, Ely, Bury St. Edmunds, Felsted, Ipswich, Allscott, Kidderminster.

the 1920s and 1930s and have helped to perpetuate this regional specialization. More recently, quick freezing, especially of peas, has affected the pattern of production, for such crops must be rapidly processed after harvesting and need to be grown close to the freezing plant. New areas of production have developed from the demands of such factories, as with the growing of French beans in east Norfolk. Figure 45 shows that the major canneries and quick-freezing plant, the main concentrations of which are in Yarmouth, Lowestoft, King's Lynn, Wisbech, Spalding, Faversham and Maidstone, are in the principal horticultural areas. Factories outside the main growing areas are primarily located in large towns.

The Extent and Composition of Agricultural Land

The extent of the land area of England and Wales in agricultural use is not known exactly, partly because of uncertainty about the area of rough land used for grazing and partly because there is no clear dividing line between agricultural and other uses, such as recreation and military training. According to the agricultural census in 1970, 28·8 million acres were used for agriculture (78% of the total area), of which 23·8 million acres, or 83%, were arable land and permanent grassland. Such land in crops and grass is usually divided into fields and receives some sort of regular treatment; most of the rough grazing is open moorland. By far the greater part of the remaining 20% of the land is accounted for by land under houses, roads and other forms of urban development (12%), and by woodland (7%).

Not surprisingly there are considerable regional differences, both in the proportion of land devoted to agriculture and in the composition of the agricultural land (Fig. 46). The lowest proportion of agricultural land is found in districts around the great conurbations, and even a part of this is probably only in semi-agricultural use as recreation grounds, accommodation fields and the like. The highest proportion occurs in the Isle of Ely and Holland, where 90% of the land is agricultural. In the uplands most other land is woodland or land which escapes enumeration; in the lowlands it is mainly urban land and woodland.

The area under rough grazings is the least well-known statistically of all the major categories of land use in the country, partly because farmers are more interested in the stock it will carry than in its extent, and partly because some of it is common land over which numbers of farmers enjoy grazing rights. Periodic estimates of the extent of common land are made by officials of the Ministry of Agriculture, but these are not claimed to have a high degree of precision; for, as the Royal Commission on Common Land discovered, there are no accurate records and some rights are not exercised. Apart from the New Forest, common rough grazings are largely confined to the uplands (Fig. 47). They total some 1·5 million acres (or 34% of the rough grazings), but the proportion varies widely. In Brecon, Westmorland and the North Riding more than half the rough grazings are common, while in Northumberland and Montgomery, counties with comparable proportions of rough grazings, less than 10%

is common. In part, these differences may simply be due to inaccurate data; a return of 1873 (whose accuracy is admittedly also suspect) estimated the commons of Montgomery at nearly 110,000 acres, compared with under 15,000 in the current agricultural returns.

There is also uncertainty about changes in the extent of land used for agriculture. In this century the trend has certainly been downward, though the loss has been minimized statistically by improvements in the accuracy of the agricultural returns.

Fig. 46. Rough grazings per 100 acres of agricultural land.

Fig. 47. Common rough grazing per 100 acres of all rough grazing.

Most of the land lost has been taken for urban development, although the Forestry Commission has been planting on moorland since 1919, while an average of some 100,000 acres of agricultural land was taken annually during the Second World War for military purposes. The return of much of this land from military use to agriculture has complicated the pattern of post-war losses, but the general trend is again downward and, in the 1960s, an average of some 57,000 acres a year were being taken out of agricultural use. Yet the seriousness of these losses of mainly above-average farmland, the equivalent of a reduction of under 5% in the total agricultural area since 1939, is diminished by the great increase in agricultural productivity; over the same period the net agricultural output of the United Kingdom more than doubled largely as a result of improvements in plant breeding, increases in the application of fertilizers to the land and other technological improvements, though a precise estimate cannot be given because of changes in the basis of calculation.

It is difficult to demonstrate the importance of most of the features described in this chapter for the production of individual crops and classes of livestock. Yet there is no doubt that they are important, and it is necessary to keep these very varied man-made conditions in mind in reading the succeeding chapters, where their importance is inevitably minimized for lack of data.

CHAPTER IV

Tillage Crops

In this chapter the distribution of crops other than grass and horticultural crops is described and analysed. Each crop comprises several varieties, with their own requirements and distributions; but detailed data are available only for the acreage under the various crops as a whole, so that it is not possible to discuss the regional differences in the varieties grown. For each of the major crops there are three maps. The first map shows the distribution of the crop, given as a percentage of the acreage under crops and grass. The other two maps show its relative importance, expressed as proportions of the tillage acreage and, where appropriate, of the cereals or roots acreage; they have been included as a measure of the competitive power of the crop in question, as compared with other crops in general and with crops of the same kind. For less important crops, the third map is omitted. In interpreting these maps, allowance must be made for the varying proportions of rough grazing (Fig. 46). All maps have been compiled from the census taken in June 1970 and all figures quoted in the text relate to June censuses unless otherwise attributed.

Tillage

Agricultural land is usually divided into arable, permanent grass and rough grazing, arable land being that which is ploughed. Owing to ambiguities about the interpretation of grass, which are considered in the next chapter, it is desirable to confine the present discussion to land under tillage, i.e., ploughed land, other than grassland, under crops and fallow. Conventionally tillage includes land under orchards, but since these can have grass, fallow or crops beneath the trees, this practice can be misleading, especially in counties such as Somerset where many orchards are grass orchards. Since the purpose of Figure 48 is to show those areas regularly ploughed, orchards have been excluded in defining and calculating the tillage acreage.

In 1970 there were 10,213,487 acres under tillage (excluding orchards), or 43% of the crops and grass acreage. Figure 48 shows the proportion of land under tillage in the A.D.A.S. districts. The highest values are found in the Fenland, where 90% or more of the crops and grass acreage is under tillage: the highest proportion is 95% in District 156 (mainly silt fen), but there is little difference within Fenland. The Fen margin and most of the remainder of East Anglia, with the exceptions of south Essex,

Fig. 48. Tillage per 100 acres of crops and grass.

73

Breckland, and Broadland, have 75% or more of their crops and grass acreage under tillage, and there are few districts in eastern England with less than half; the chief exception is south-east England, especially the Low Weald. Outside eastern England, the only other area where tillage accounts for 50% or more is the chalk downland of Berkshire, Hampshire and Wiltshire, although there are also outliers in south-west Lancashire and the west Midlands. Apart from these, the proportion of tillage in western districts is less than 50%, while over most of the uplands of Wales and northern England, the proportion is under a tenth, the lowest values being found in District 80, where less than 1 acre in every 1,000 of crops and grass is used for tillage.

Inevitably a map based on district data gives an imperfect idea of the detailed distribution of land under tillage. Thus in Norfolk much of the land not under tillage is to be found in river valleys and on alluvium, so that the tillage acreage could not easily be increased without expensive drainage works; in the Wealden counties, there is a sharp contrast between proportions on sand and loam and on clay.

The area under tillage changed markedly over the previous 32 years. In 1938 there were only 6,725,079 acres in England and Wales (27% of the crops and grass acreage); the acreage increased rapidly during the Second World War, reaching a peak of 11,331,450 acres (47%) in 1944. The expansion of this tillage acreage was achieved partly by an increase in eastern districts, but there was also a very large increase in western counties and on heavy land in the Midlands. In Wales, for example, the tillage acreage rose from 287,712 acres in 1939 (11%), to 880,722 acres in 1943 (35%). From the end of the War the acreage declined until the late 1950s, when it began to rise steadily again, first in eastern England, where there is now more land under tillage than during the war-time peak, and then in Midland counties. In Wales and some western counties of England it continued to decline; in Wales, for example, there were only 281,076 acres in 1970, a smaller acreage even than in 1939. Nearly everywhere else the acreage under tillage is higher than in the 1930s, though the contrast between east and west is now greater. Such expansions and contractions have also been a feature of most crop distributions during the past 30 years, the chief exception being barley, of which the acreage increased steadily to make it the leading crop nearly everywhere.

In a broad view, controls over the distribution of tillage are primarily climatic, the main tillage areas having less than 30 inches of rain per annum and relatively few rain days (cf. Figs. 3 & 4), while the lower probability of good harvest weather restricts the acreage sown to most crops in the western half of the country. Districts with the lowest proportion of the crops and grass acreage under tillage are in areas of high rainfall, particularly in the north-west where lower accumulated temperature and fewer hours of sunshine further reduce the prospect of a successful harvest. Western districts with above-average proportions of tillage are found either in the south-west, where temperatures and sunshine are more favourable for crop growth, or around the coast, where frost hazards are reduced and there are possibilities of growing early crops (as in south Pembrokeshire), or in areas such as south-west Lancashire, the Welsh borderland and the Eden valley, which lie in the rain-shadow of the uplands.

The predominance of tillage in eastern districts is not wholly due to the advantages of their climate. The drier climate is said to make the establishment of grass more difficult, while periods of drought in summer can severely restrict grass growth. Their unsuitability for grass is thus also a factor, though it is much less important now than formerly.

High elevation and deeply dissected terrain, acting both directly and through modifications of climate, greatly restrict the areas suitable for tillage in the west and north of the country. But within the eastern areas suited to tillage on climatic ground, the nature of the soil and the height of the water-table play an important part in determining what land is used for crops other than grass. The pre-eminence of eastern England is reinforced by the fact that most of the land best suited to arable cultivation lies in eastern districts (Fig. 10). Soils of light and medium texture, which can be worked at most times of the year and are well-drained, are favoured, especially where they lie on gently undulating terrain. Many of the soils of East Anglia are of this kind, notably the silts and peats of the Fenland and the loams of central Norfolk; soils in High Norfolk, Suffolk and Essex are heavier but are generally calcareous and do not present major problems of drainage. Soils on the drift-free chalk of Berkshire, Wiltshire and Hampshire, though shallow, are generally well-suited to mechanized corn growing, while the extensive use of fertilizers, the development of new varieties of crops and the profitability of arable farming in the post-war period have greatly increased their importance. Low proportions of tillage in eastern England are generally associated with heavy soils and impeded drainage. Most western districts with above-average proportions of tillage similarly have lighter soils, particularly those derived from peat and sand, as in south-east Lancashire, and from sandstone, as in the Eden valley and the southern part of the Plain of Hereford. Many of the soils of the south-west are also medium or light in texture, the chief exceptions being those developed on the Culm Measures in west Devon.

Reinforcing these physical causes are considerations of farm size, farm equipment and the nature and availability of farm labour. Many occupiers in the principal tillage areas have large farms and fields and have invested heavily in tillage machinery; they are thus unlikely to turn to grass farming, even if arable farming becomes less profitable. Similarly the existence of a skilled labour force for the growing of intensive crops such as potatoes and vegetables is an added incentive to retain land in tillage, while workers accustomed to piece-work rates and the 5-day week are unlikely to exchange these willingly for the 7-day week of the stockman.

Cereals

The major role which cereals play in the cropping of land under tillage is partly due to their importance in providing cash crops, feed crops (both as grain and straw) and bedding. They also lend themselves to mechanization and require little attention between sowing and harvest, both relevant considerations at a time when labour is becoming increasingly scarce and costly. Furthermore, it is said that, in the main areas of crop farming, it is difficult to devise satisfactory rotations which do not include a large acreage of cereals.

In 1970 there were 7,859,235 acres under cereals, or 77% of the tillage acreage. Figure 49 shows the proportion of the tillage acreage in cereals throughout the country. In most districts it exceeds 70%, the chief exceptions being those areas where vegetables and cash roots are important crops, notably the Fenland (although even here more than 40% is under cereals) and the uplands of the west and north.

Fig. 49. Cereals per 100 acres of tillage.

Fig. 50. Leading cereals.

Proportions of 85% and over are to be found in the south Midlands and on the chalklands of southern England, District 212, with 93%, having the highest percentage. Districts with under 60% of the tillage acreage under cereals are to be found in two quite dissimilar parts of the country, viz., in areas of intensive arable farming, as in East Anglia, and in areas where there is little land under tillage crops, as in Wales and north-west England.

While individual cereals differ greatly in their requirements, cereals can be grown under a wide range of conditions. The major controls are climatic, particularly the possibility of harvesting ripened grain, which diminishes northward as the growing season begins later and westward as rainfall and number of rain-days increase. For

76

most cereals, yields and quality are also higher in eastern parts of the country. Cereals are generally less demanding in their soil requirements than roots, but are favoured by both natural and man-made conditions suitable for mechanized farming, as on the chalklands of Wiltshire. But many of the differences shown in Figure 49 can also be explained by the comparative advantages of other crops over cereals; thus proportions of cereals are low in eastern and southern England where the land is suitable for sugar-beet, potatoes and vegetables, while in the west and north-west fodder brassicas and roots, such as rape and turnips, which are better suited to the prevailing climatic conditions, tend to displace cereals.

The distribution of the individual cereals is analysed in detail in succeeding pages, but Figure 50 indicates the leading cereal in each district. The map is dominated by the symbol representing barley, which is the leading cereal in 90% of the districts. The chief exceptions are the Fenland and parts of the west Midlands, where wheat is the first choice; central Wales and parts of the Pennines, where oats still retain their primacy; and west Pembrokeshire and Cornwall, where mixed corn is the chief cereal. This pattern is in marked contrast to that in the first edition of this atlas, when wheat, barley and oats were almost equally important.

The acreage under cereals has, of course, fluctuated with the tillage acreage over the last three decades, but the proportion has tended to increase at the expense of fodder root crops, particularly in southern and midland counties. In 1938 there were 4,127,834 acres under cereals, or 61% of the tillage acreage, and in 1943 this had increased to 7,634,008 acres (68%).

Wheat

Wheat is the principal cereal grown for human consumption, although it is also an important source of livestock feed. In 1970 there were 2,394,215 acres under the crop, or 10% of the crops and grass acreage, 23% of the tillage acreage and 30% of the acreage under cereals. The principal wheat growing areas are in eastern counties, chiefly in Lincolnshire, Essex, Huntingdonshire and Cambridgeshire, all of which have more than 20 acres of wheat in every 100 acres of crops and grass (Fig. 51). The highest proportions are found in the southern Fenland, District 190 having 32%. South-west Lancashire is the only western district where wheat is an important crop, for elsewhere in western England and in Wales less than 5% of the crops and grass acreage is under the crop. The 5-year average yield (1966–7 to 1970–1) was 31·4 cwts./acre. High yields are obtained mainly in eastern counties, the highest being 36·7 cwts./acre in Northumberland.

This distribution shows both the climatic and edaphic preferences of wheat. Most of the wheat is grown where there is less than 40 inches of rain per annum because of the difficulty of ripening and harvesting the crop elsewhere, and because of the lower yields and its intolerance of soil acidity in high-rainfall areas. Within the climatically favoured areas, wheat, with its strong rooting system, is widely grown on heavy to medium textured soils.

Fig. 51. Wheat per 100 acres of crops and grass.

While wheat accounts for one-quarter or more of the tillage acreage in nearly all the principal wheat growing areas, the highest proportions are to be found in the south Midlands, where a third or more of the tillage acreage is sown to wheat (Fig. 52), though the highest value is 41% in District 209. These high proportions arise from the prevalence of heavy soils in those districts and the limited range of crops which can be grown on them.

The map showing the proportion of the cereal acreage under wheat presents a somewhat different picture (Fig. 53). In and around the Fenland, in east Lincolnshire, Essex and part of the south Midlands, the proportion of wheat exceeds 40% and is

Fig. 52. Wheat per 100 acres of tillage.

Fig. 53. Wheat per 100 acres of cereals.

highest in the Fenland, where some 60% of the cereal acreage is under wheat; in District 154 a value of 74% is recorded. In this particular respect, as in many aspects of its agriculture, the Fenland is unique. In south-west and north-east England and in Wales, by contrast, not only is the acreage under wheat very restricted, but the crop accounts for only a small proportion of both tillage and cereal acreages.

The acreage under wheat rose from 1,830,272 acres in 1938, or 7% of the crops and grass acreage, 27% of the tillage acreage and 44% of the cereal acreage, to 3,279,905 acres in 1943 (13%, 29% and 43% respectively), largely because of the need to expand the production of home grown wheat during the wartime blockade. Since 1945 the acreage has remained fairly stable at about 2 million acres, although the crop was replaced as principal cereal by barley in the mid-1950s. Because of the extension of

79

land under tillage, the share of the wheat acreage grown in the main arable counties of eastern England fell, though they remain the principal wheat-growing areas. The acreage under wheat is restricted by its susceptibility to fungal diseases if grown too frequently and by the unsuitability of most home-grown wheat for bread-making.

Barley

Barley is likewise an important cash crop, being sold both for brewing and distilling and as feed for livestock; it is also widely consumed by livestock on the farms of origin. In 1970 there were 4,710,338 acres under barley, or 20% of the crops and grass acreage, 47% of the tillage acreage, and 60% of the acreage under cereals. The main areas for barley growing are in eastern counties, in East Anglia and the East Riding, and also in Hampshire and Wiltshire. Only the Fenland, where the crop does not do well, stands out as anomalous among the main tillage areas of eastern England (Fig. 54). The most important area is Norfolk and Suffolk, in several districts of which more than 30% of the crops and grass acreage is under barley, but the highest proportion is 48% in District 26. In most of Wales and north-west England, barley occupies less than 5% of the comparatively small acreage under crops and grass. The national 5-year average yield was 27·9 cwts./acre; high yields are obtained in northern and west midland counties, the highest average being 33·7 cwts. in Westmorland.

Although barley can be grown under a wide range of conditions, this distribution accords with its preference for light, calcareous or non-acid soils; soil acidity is the most common cause of crop failure. Many farmers hope to make a malting sample and suitability for malting decreases to west and north, although it also depends greatly on farming skill; but malting quality is now much less important than it was when a higher proportion of the crop was sold for malting and the demands of the maltsters dictated the price. Stiff-strawed varieties of barley are both resistant to lodging and particularly suited to combine harvesting. Barley is also resistant to many of the diseases that beset wheat and oats, and on the chalklands as many as fifteen crops of barley have been taken in succession. Although the crop is grown on farms of all sizes, large barley farms are characteristic of the chalklands, and a third of the crop is grown on farms of 500 acres and over, or nearly double their proportionate share. The more northerly extension of the crop compared with wheat is also a reflection of its shorter growing period.

A high proportion of the tillage acreage is under barley in two contrasting areas, on the chalk and limestone uplands of southern England, which are well suited to barley but not to other cash crops which compete where soils are deeper, as in East Anglia, and northern England, where most crops are grown for fodder and barley has largely replaced oats in recent years (Fig. 55). In both these areas, more than half the tillage acreage is under barley, the highest value being 71% in District 48. In Wales and the Fenland, barley occupies less than a third of the tillage acreage. Northern England and the chalklands are also prominent on the map showing the proportion of the cereal acreage under barley (Fig. 56); the leading areas for wheat have relatively low

Fig. 54. Barley per 100 acres of crops and grass.

proportions under barley, as do those where oats and mixed corn are still popular.

There were only 885,499 acres under barley in 1938, or 4% of the crops and grass acreage, 13% of the tillage and 22% of the cereal acreages. The acreage rose to a maximum of 2,002,890 acres in 1946 (8%, 19% and 30% respectively), but, after a brief fall, it has shown an upward trend throughout the 1950s and for much of the 1960s. East Anglia has long been and remains the chief region for barley, but the crop is now more widely grown and the share of eastern counties has fallen (although their acreage has increased considerably). This spread has largely been at the expense of oats and mixed corn and of roots and other fodder crops. It is due mainly to the

Fig. 55. Barley per 100 acres of tillage.

Fig. 56. Barley per 100 acres of cereals.

development of higher-yielding, short-strawed varieties suitable for mechanical harvesting, to its greater popularity as a feed grain, and to improvements in soil status as a result of widespread and subsidized liming.

Oats

Oats are used mainly for feeding to livestock, particularly horses and cattle, being valued both as a grain and for the high quality of the straw; little of the crop leaves the farms on which it is grown and the proportion used for human consumption is small. There were 574,933 acres under the crop in 1970, or 2% of the crops and grass acreage, 6% of the tillage acreage and 7% of the cereal acreage. Oats are widely grown but their distribution has become increasingly patchy as the acreage under oats has declined (Fig. 57). The highest value is only 6% of the acreage under crops and grass

Fig. 57. Oats per 100 acres of crops and grass.

in District 23, and it is easier to identify those areas where proportions are low (the Pennines and the Fenland), than to summarize those where oats are relatively important; subsequent maps present a more coherent picture (Figs. 58, 59). The national 5-year average yield was 28·6 cwts. per acre. High yields are obtained in both western and eastern counties, the highest being 32·7 cwts. in Wiltshire.

The relative importance of oats is most marked in western districts, especially Wales and north-west England where a fifth or more of admittedly small acreages under tillage are devoted to oats (Fig. 58); the highest proportion is 41% in District 322. In the principal cereal areas of eastern and southern England, oats account for

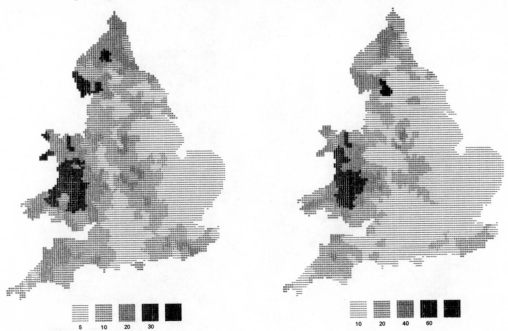

Fig. 58. Oats per 100 acres of tillage.   *Fig. 59. Oats per 100 acres of cereals.*

less than 5% of the tillage. The map showing the proportion of the cereal acreage devoted to oats is very similar, with highest values in western districts, notably central Wales (Fig. 59), reaching 95% in District 328, whereas in eastern districts the proportion is less than a tenth. The areas where oats were the leading cereal were much more widespread a decade earlier.

This distribution reflects both the ecological characteristics of the crop and also the strong competition it meets from wheat and barley in eastern and southern districts. Oats are more tolerant of acid soils than either wheat or barley, grow well under moist and cool climatic conditions and ripen with a minimum of sunshine, while the crop is liable to fail in dry seasons in eastern counties. Oats also have the advantage that they

Fig. 60. Mixed corn per 100 acres of crops and grass.

can be harvested green as well as being ripened for grain. Oats are at a disadvantage compared with other cereals in the main arable areas because they lend themselves less readily to combine harvesting and because yields and returns per acre are lower. It must not, however, be assumed that conditions are ideal in the west and north, for costs of production there are often high and harvesting difficult.

There were 1,300,738 acres under oats in 1938, or 5% of the acreage under crops and grass, 19% of the tillage acreage and 32% of the cereal acreage. The crop was widely grown in the first few years of the Second World War, the maximum acreage of 2,499,384 acres being reached in 1942 (10%, 24% and 37% respectively); since then

Fig. 61. *Mixed corn per 100 acres of tillage.*

Fig. 62. *Mixed corn per 100 acres of cereals.*

the acreage has declined fairly steadily, partly owing to the decline in the number of horses, for which oats are staple feed, but chiefly owing to competition from higher-yielding barleys. Such competition has even extended to the upland margins and only in a few areas do oats survive as the leading cereal.

Mixed Corn

Mixed corn, that is, mixtures of wheat, barley, oats, beans, peas or vetches, is grown for feeding to livestock, but only the acreage grown for harvesting as corn is now separately enumerated. In 1970 there were only 164,809 acres under mixed corn for threshing, 0·7% of the crops and grass acreage, 1·6% of the tillage acreage and 2·1%

of the cereal acreage. West Cornwall has the largest acreage under mixed corn, and in some districts 10% or more of the crops and grass acreage is under the crop, the highest proportion being 12% in District 296 (Fig. 60). Other important areas are in south-west Wales and the north-west Midlands, especially Cheshire. The crop plays only a minor role in eastern England. The 5-year national average yield is 27·1 cwts./acre, and yields are highest in eastern and northern counties, with the highest average in Northumberland (32·3 cwts.).

It is not clear why the crop should be favoured in the south-west peninsula, where the normal mixture is one of barley and oats, but the leading position of Cornwall has certainly been long established. It has been held that higher yields were obtainable from mixed corn than from the component crops (though experimental evidence in Cornwall does not generally support this view) and that mixed corn has been grown as insurance against the failure of either crop. A possible factor in this regional concentration may be the practice of dressing the land with calcareous sand which reduces soil acidity. Whatever the original reasons, the growing of the crop now seems to be due largely to the influence of tradition, and it is rapidly being replaced by barley.

The leading position of Cornwall is clear from whatever viewpoint the crop is considered. Mixed corn accounts for 30% of the tillage acreage in west Cornwall, the highest proportion being 40% in District 298 (Fig. 61); the crop also accounts for over 40% of the cereal acreage in this area, the proportion reaching 71% in District 44 (Fig. 62). In parts of Wales, the Lake District and north-west Midlands mixed corn is a relatively important crop, accounting for 10% of both tillage and cereal acreages, whereas in eastern and southern England proportions are nearly everywhere under 1%.

There were 92,853 acres under mixed corn in 1938, or 0·4% of the acreage under crops and grass, 1·4% of the tillage acreage and 2·2% of the cereal acreage, nearly half of it being in Cornwall. The acreage rose during and after the Second World War, when feeding-stuffs were rationed and the feeding of wheat and barley to livestock was forbidden, to a maximum of 826,929 acres in 1950 (3·4%, 8·1% and 12·2% respectively). Since then the crop has been gradually displaced nearly everywhere by high yielding barleys, although, owing to the fact that since 1955 only the acreage for threshing has been separately recorded, trends cannot be accurately determined. On the basis of these rather different figures, the relative importance of Devon and Cornwall has declined considerably, their share of the total acreage falling from 57% in 1939 to 31% in 1970.

Rye

The small acreage of rye for threshing (only 11,397 acres in 1970, or less than 0·1% of the crops and grass acreage, 0·1% of the tillage acreage and 0·1% of the cereal acreage), is in marked contrast to the situation in many other European countries in similar latitudes, and although 118,241 acres were grown in 1943 (when rye for threshing was first recorded separately), rye has long ceased to be a major crop. Over most parts of the country a very small acreage is grown, and the distribution is highly fragmented on both maps (Figs. 63, 64). It does, however, show some association with areas of light soils, notably the Breckland and the Suffolk Sandlings. The highest values are 2% of crops and grass in District 159 and 3% of tillage in District 178. This distribution arises, not because rye prefers such soils, but because it can tolerate them and because of the greater comparative advantage of other cereals elsewhere.

Maize

Maize is a recent introduction as a field crop on British farms and was first recorded in the agricultural census in 1970. A total of 3,543 acres was returned in that year, of

Fig. 63. Rye per 10,000 acres of crops and grass.

Fig. 64. Rye per 1,000 acres of tillage.

which 1,060 acres represented maize being grown for threshing. Understandably, in view of its origins, the crop is largely confined to east and south-east England, notably along the southern coast of East Anglia, although isolated high values occur elsewhere (Fig. 65, 66). The highest values recorded are 0·4% of the crops and grass acreage and 0·7% of the tillage acreage in District 244.

Root Crops

Although they are not all strictly root crops, the following crops will be considered as such for present purposes: fodder-beet, mangolds, potatoes, sugar-beet, turnips and swedes. These crops, the harvested parts of which grow in or on the soil, have a number of features in common; they are bulky, have heavy labour requirements and are grown, for the most part, in relatively small acreages. They include both important cash crops and succulent fodder crops, which are a useful complement to cereals. Because of the intensive cultivation which they require, root crops are also valued as cleaning crops, although the development of chemical sprays for weed control has made them much less important than formerly.

There were 1,093,204 acres under roots as here defined in 1970, or 11% of the tillage acreage. The regional variation in the importance of root crops is more marked than in the case of cereals (Fig. 67). The Fenland, with large acreages under sugar-beet

Fig. 65. Maize per 10,000 acres of crops and grass.

Fig. 66. Maize per 1,000 acres of tillage.

and potatoes, is pre-eminent, with 20% or more of the tillage acreage under roots, District 190 having the highest percentage (35%). Other areas with high proportions are the Vale of York, east Shropshire and south Staffordshire. Districts with a fifth or more of the tillage acreage under roots are found in two quite different locations: on the one hand, in Norfolk, the Fen margins, the Humber warplands, the Vale of York, the Midlands and south-west Lancashire, which all have large acreages under tillage, and on the other, in North Devon, Brecon, Radnor and north-west England, where there is little tillage.

Individual root crops differ considerably in their requirements, some being best suited to the moist, cool climate of the north-west and others to the relatively dry and sunny south and east. The major controls over the distribution of root crops are soils and drainage. Most roots require a fine seed-bed and are favoured by a deep, well-drained soil of light to medium texture, which is free of stones, for these make for ill-formed crops and difficult harvesting. Shallow, stony and heavy, ill-drained soils are generally avoided, though a small acreage of roots is grown on many farms.

The different root crops predominate in various parts of the country and the pattern has changed little since 1958, unlike that of cereals (Fig. 68). Potatoes are the leading root crop throughout most of midland and southern England and around the margins of the uplands. Turnips and swedes are the principal root crop in northern England, the Pennines, east Yorkshire, upland Wales and Devon, though the acreages are often

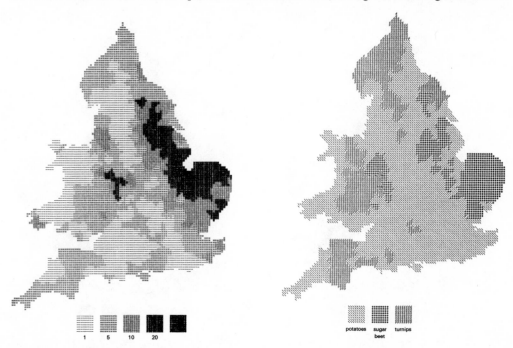

Fig. 67. *Root crops per 100 acres of tillage.*

Fig. 68. *Leading root crops.*

small. Sugar-beet is the leading root crop throughout East Anglia, except the Fenland, in much of the east Midlands and in a few districts in the west Midlands.

The acreage under root crops has fluctuated considerably over the previous 32 years, while the proportion of the tillage acreage under toots has tended to decline. In 1938 there were 1,423,374 acres under roots (21% of the tillage acreage), compared with 2,168,845 acres (19%) in 1944 and 11% in 1970. This decline has been fairly general, but has been least marked in those districts where large acreages of cash roots are grown; it is partly due to high labour costs, to the use of chemical sprays and to the expansion of the acreage under kale, which has replaced roots as a fodder crop in many areas.

Fig. 69. Potatoes per 1,000 acres of crops and grass.

Potatoes

Potatoes are grown primarily as a cash crop, but because the yield of potatoes varies considerably from year to year while the demand for potatoes changes little, a variable proportion of the marketable crop is surplus to requirements and is sold for processing or as stock feed, while substandard potatoes are fed to livestock on farms. Even the marketable crop comprises three separate crops, viz., seed potatoes, early potatoes and main crop potatoes, although the last accounts for the greater part of total acreage. Eleven per cent of the acreage planted in 1970 was under early potatoes; the acreage of certified seed potatoes (not separately distinguished in the agricultural returns) was

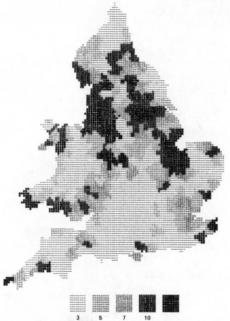

Fig. 70. Potatoes per 100 acres of tillage.

Fig. 71. Potatoes per 100 acres of root crops.

8,642, or less than 2% of the acreage grown by producers registered with the Potato Marketing Board.

There were 512,629 acres of potatoes grown in 1970, or 2·2% of the crops and grass acreage, 5·0% of the tillage acreage and 47% of the acreage under root crops. The principal potato-growing area is the Fenland, where 10% or more of the crops and grass acreage is under potatoes, the highest proportion being 18% in District 190 (Fig. 69). Other major areas are the Fen margins, the Humber warplands, south-west Lancashire, south Staffordshire and mid-Bedfordshire. Acreages are small in the uplands, East Anglia and the southern chalklands, while isolated western districts, such as south-west Pembrokeshire and west Cornwall, are associated with early potato production. The small acreage grown for seed is to be found mainly in western and northern districts. The 5-year average national yield is 10·8 tons/acre. In general the highest yields are to be found in counties in eastern England and along the Welsh border, although the highest average yield recorded in 1966–7 to 1970–1 is 12·7 tons/acre in west Sussex.

While potatoes will grow under a wide range of conditions, they do best in deep, well-drained and stone-free soils of light to medium texture, such as the silts of the Fenland and the soils of the Humber warpland. The leading position of the northern Fenland in potato growing was established in the 19th century and, while there is no evidence that costs are lower than elsewhere, yields are appreciably higher; furthermore, potatoes grown in such soils command a premium on account of their quality. By contrast, potatoes from the peats of the southern Fenland are of poor quality, though yields are high. Another advantage of the Fenland is the availability of casual labour, for the potato harvest is reputed to occupy half a million men, women and children for 10 days each autumn. Markets are also important, although the main producing area is not near any of the major markets, but roughly equidistant between them. Producers in Yorkshire, Lincolnshire, north Cheshire, Staffordshire, Kent and Essex are conveniently placed for access to urban markets, and the importance of markets in the north is enhanced by the higher consumption of potatoes per head in northern counties. The location of potato growing is also affected by the activities of the Potato Marketing Board, with which all producers of one acre or more of potatoes must register. Each grower is allocated a basic acreage (which is modified from time to time), and the acreage he grows may be restricted by decision of the Board, fines being levied on any excess acreage. In 1970, the quota was 95% of basic acreages.

Although potatoes are little grown on heavy, stony or shallow soils, heavy eel-worm infestation in the Fenland has encouraged growers to look elsewhere and there has been some increase of production on heavy land, notably in Essex. The crawler tractor has facilitated production on heavier soils, although lifting the crop often presents difficulties. It must not be assumed that the present distribution reflects accurately the suitability of different soils. Apart from the general effects of controls on production, competition from sugar-beet affects the acreage grown in some areas, such as East Anglia.

The relative importance of potatoes is also greatest in the Fens, in several districts of which the proportion of the tillage acreage under potatoes is 20% or more; reaching 29% in District 352 (Fig. 70). The prevalence of eel-worm infestation in this area suggests that the proportion could not easily be increased. In south Pembrokeshire, too, a fifth of the tillage acreage is under potatoes, mainly earlies. Figure 71, showing the proportion of the root acreage under potatoes, presents a somewhat different picture. The relative importance of potatoes in districts near the main urban markets, where they occupy over 75% of the root acreage (99·5% in District 80), is greater than the actual acreage would suggest. In the Fenland and the other principal potato-growing areas the proportion of potatoes is reduced by the large acreages devoted to sugar-beet.

In 1938 there were 474,786 acres under potatoes, or 1·9% of the crops and grass acreage, 7·1% of the tillage acreage and 34% of the acreage under root crops. The acreage rose during and after the Second World War to meet food shortages, reaching a maximum of 1,116,976 acres in 1948. Since then demand has fallen and, with yields rising, the acreage has been reduced, the principal contraction occurring in those areas to which potato growing had spread during the war. The chief exception is Pembrokeshire, where more than six times the pre-war acreage was grown in 1970. In Essex, too, the acreage has doubled. Nevertheless, the midland counties generally account for a larger proportion of the total acreage than before the war, while both the relative and absolute importance of the Fenland have declined. The acreage grown in Holland in 1970 was only half that in 1939 and the country's share of the total acreage has been almost halved.

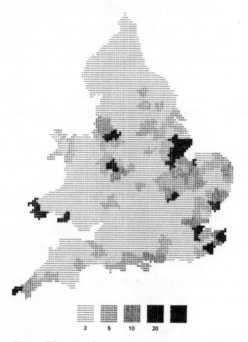

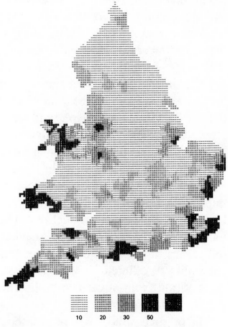

Fig. 72. Early potatoes per 1,000 acres of crops and grass.

Fig. 73. Early potatoes per 100 acres of potatoes.

Early Potatoes

A total of 66,425 acres was planted in 1970 for harvesting before 31 July. They are grown both in the main areas for potatoes and in districts specializing in early potatoes; but, apart from some districts in the west Midlands and Bedfordshire, most of the areas growing early potatoes are coastal (Fig. 72). The earliest potatoes are grown in south Pembrokeshire, west Cornwall and Kent, and may be lifted in May, but the biggest acreage comes from the Fenland; the highest proportion of the crops and grass acreage (7%) was recorded in District 352. Yields are only two-thirds of those for main crop potatoes, but high prices are obtained for crops marketed sufficiently early. The relative importance of earlies is generally more marked in southern coastal districts, although only in Pembrokeshire, in west Cornwall and in north-west Wales do first earlies account for more than half the potato acreage, the proportion reaching 91% in District 299 (Fig. 73), compared with the national average of 12%. The early onset of the growing season and freedom from frosts are clearly the relevant factors, although the actual location of production often depends on local conditions of site and soil. Some of the land used for growing early potatoes in Cornwall and Pembrokeshire is not ideal for cultivation, but is sheltered and has a favourable aspect. Early potatoes recorded in June 1938 totalled 58,498 acres, or 12% of the potato acreage. Production is now more widespread. Whereas the Fenland counties contained 24% of the total acreage in 1939 and Lancashire and Cheshire 19%, these proportions had fallen to 9% and 8% respectively in 1970, while the share of Cornwall and Pembrokeshire rose from 4% to 15% (though figures are not strictly comparable).

Sugar-beet
While sugar-beet is primarily grown for the extraction of sugar, it is also valuable as a source of fodder; thus the tops and pulp from an acre of beet are said to equal the feeding value of an acre of oats. Sugar-beet is grown only under contract for the British Sugar Corporation, and the total acreage which may be grown is determined by the Government in the light of commitments under the Commonwealth Sugar Agreement. The growing of the crop was originally encouraged by Government action to alleviate the effects of the depression in arable farming in the 1920s.

There were 450,060 acres of sugar-beet grown in 1970, or 1·9% of the acreage under crops and grass, 4·4% of the tillage acreage and 41% of the acreage under roots. Nearly 90% of the sugar-beet is grown in eastern England between the Thames and the Vale of York; only south Essex, Holderness and east Lincolnshire have less than 10 acres per 1,000 acres of crops and grass under the crop (Fig. 74). The main areas of production are in Norfolk, west Suffolk and the peat Fenland, where the crop accounts for more than 10% of the land under crops and grass, the highest percentage being 19% in District 187. There is one outlier of lesser importance in Shropshire, Worcester and Herefordshire. Small acreages are grown throughout midland and southern England and the crop is absent from a number of districts in northern England and upland Wales. The estimated average yield in the 5-year period 1966–7 to 1970–1 for beet handled by the British Sugar Corporation was 14·9 tons. No county data on yields are available, but the Corporation has provided yields for its seventeen factories in England and Wales (cf. Fig. 43); on the reasonable assumption that most of the acreage under contract to a given factory lies fairly close to it, these figures show that higher yields are obtained in East Anglia and the Fenland, the highest being 16·7 tons for the Spalding Factory.

Fig. 74. Sugar-beet per 1,000 acres of crops and grass.

This distribution is related partly to the physical requirements of the crop, partly to the location of the sugar factories and partly to Government policy. Sugar-beet does not do well in most western and northern areas and yields there are low. Within the drier and warmer areas, the location of sugar-beet growing is governed chiefly by soil requirements. The ideal soil is a deep, stone-free loam of high base-status, but in practice soils range from sands and peats to loams and heavy loams; shallow, stony or heavy, poorly-drained soils are avoided. In wet seasons heavy soils are at a disadvantage, for not only is harvesting difficult, but considerable quantities of soil are removed with the beet; at the Felsted factory, for example, which draws most of its

Fig. 75. Sugar-beet per 100 acres of tillage.

Fig. 76. Sugar-beet per 100 acres of root crops.

beet from the heavy loams of Essex, dirt accounted for 23% of the weight delivered at the factory in 1961, compared with 12% from Fenland beet. Although planting and harvesting have been successfully mechanized, the crop still makes heavy demands on labour during the period of growth.

The location of sugar-beet growing is also influenced by the siting of factories; with the exception of that at Cantley, erected in 1912, these were all built in the 1920s by commercial companies and were presumably located in areas thought suitable for the growth of the crop. Whatever the reason for their location, Government unwillingness to permit the erection of further factories elsewhere, together with limitations on the total acreage grown, has ended to fossilize production in areas near those factories.

Fig. 77. Turnips, swedes and fodder-beet for stock feed per 1,000 acres of crops and grass.

But it has also been policy to spread the benefits of sugar-beet growing more widely by permitting production of the crop in counties away from the main growing areas.

The maps showing the distribution of areas with a high proportion of the tillage and root acreages under sugar-beet are very similar (Figs. 75 and 76). In most districts in Norfolk and Suffolk over 10 acres per 100 acres of tillage are under sugar-beet, the highest proportion being 21% in District 187. Similarly, sugar-beet accounts for more than half the root acreage throughout Norfolk and Suffolk, as it does in all main growing areas, except where, as in the Fenland, competition from potatoes reduces the proportion. The highest value, 92% of land under roots, occurs in District 175.

The acreage under sugar-beet rose rapidly from negligible proportions in the early 1920s, following the passing of the Sugar Industry (Subsidy) Act of 1925. In 1938 there were 328,645 acres, or 1·3% of the crops and grass acreage, 4·9% of the tillage acreage and 23% of the acreage under roots. The acreage increased during the war, but has since remained fairly steady at a little over 400,000 acres. For reasons already noted, the distribution of the crop has changed little, although the share of Norfolk, Suffolk and the Isle of Ely has fallen.

Turnips, Swedes and Fodder-beet for Stockfeeding
Turnips, swedes and fodder-beet (not separately distinguished in the agricultural returns) are fodder crops which have declined greatly in importance over the past century. The acreage mapped here is only that used for stockfeeding. Some turnips and swedes are also grown for human consumption (Fig. 132), and a proportion of the acreage grown as a catch crop probably escapes enumeration.

The acreage recorded as under turnips, swedes and fodder-beet in 1970 was 106,491, or 0·4% of the crops and grass acreage, 1·0% of the tillage acreage and 10% of the acreage under roots. The main growing areas are in western and northern districts, the leading areas being east Yorkshire and north Northumberland (Fig. 77), where the crop occupies 3 or more acres out of every 100 under crops and grass; the highest proportion is found in District 26 (4%). The national 5-year average yield is 17·8 tons/acre. High yields are obtained mainly in northern counties, the highest being in Northumberland, with an average of 23·5 tons.

This distribution is related to climatic conditions favourable to turnips and swedes and to their use for sheep feed. They thrive under cool, moist conditions, preferably on light soils, and are more liable to mildew in drier areas. They are important fodder crops in some of the main sheep and cattle rearing areas, and were formerly widely grown on the light soils of the main arable areas as part of the Norfolk 4-course rotation (wheat, turnips, barley, clover), although climatic conditions were not very favourable for the crop. With the virtual disappearance of the folded sheep, these crops are no longer so necessary and, in any case, are handicapped by their fairly heavy labour requirements and by competition from cash root crops, kale and cereals; but they are still grown to fill the gap in winter feed between late and early-growing varieties of grass. Fodder-beet has a more southerly distribution.

99

The importance of western and north-western districts is more marked when the proportions of the tillage and root acreages occupied by these crops are considered. In north-west England, Yorkshire, upland districts of south Wales and north Devon, they account for 10% or more of the tillage acreage, the highest proportions being 43% in District 43 (Fig. 78). In these same areas and throughout upland Wales and Devon, over 60% of the root acreage is under these crops, the highest value being almost 100% in District 44 (Fig. 79). By contrast, in most lowland areas, less than 1% of the tillage acreage or 5% of the acreage under roots are devoted to turnips, swedes and fodder-beet.

Fig. 78. Turnips, swedes and fodder-beet for stock feed per 100 acres of tillage.

Fig. 79. Turnips, swedes and fodder-beet for stock feed per 100 acres of root crops.

In 1938, when fodder-beet was not separately enumerated, turnips and swedes occupied 406,461 acres, or 1·6% of the crops and grass acreage, 6% of the tillage acreage and 29% of the acreage under roots. The acreage increased slowly to a maximum of 506,347 acres in 1942 and the crops have since declined steadily in acreage and in importance. Production has been increasingly concentrated in northern counties, while the arable counties' share of a declining acreage has fallen (although, owing to the addition of fodder-beet, acreages are not strictly comparable).

Mangolds
The mangold, a succulent crop related to the sugar-beet and used for livestock feeding, is also of declining importance. Like turnips and swedes, it is valuable as a useful

100

Fig. 80. Mangolds per 10,000 acres of crops and grass.

complement to hay and straw, and is used primarily as winter feed for cattle, being particularly suited to dairy cattle. A total of 24,024 acres was grown in 1970, or 0·1% of the crops and grass acreage, 0·2% of the tillage acreage and 2% of the acreage under roots. The main areas of production are in east Norfolk and in the Vale of York, the highest proportion being 1·0% in District 164 (Fig. 80). Other areas of lesser importance are Lincolnshire, Suffolk, the Welsh Borderland and the south-west peninsula. The national 5-year average yield is 24·7 tons per acre. With the exception of the Fenland, yields are highest in western and south-western counties, Cornwall, with 32·4 tons, recording the highest average.

Fig. 81. *Mangolds per 1,000 acres of tillage.*

Fig. 82. *Mangolds per 100 acres of root crops.*

Unlike the turnip, the mangold is suited to warmer and sunnier areas and to heavier soils. For these reasons, and because it outyields turnips and swedes, it is preferred in southern districts, although it is increasingly common for small acreages to be grown on farms in northern districts.

Maps of the percentage of the tillage and root acreages under mangolds resemble a map of yields, in that the highest values are found in the south-west, rather than in areas where the largest acreage is grown. In parts of the Welsh Borderland and in Somerset and Devon mangolds account for 1% of the tillage acreage, the highest proportion being 2% in District 330 (Fig. 81). Over much of the south-west a tenth or more of the root acreage is under mangolds, the proportion reaching 28% in District 279A (Fig. 82).

Tillage Crops

The acreage under mangolds rose from 213,482 acres in 1938 or 0·9% of the crops and grass acreage, 3·2% of the tillage acreage and 15% of the acreage under roots, to 300,490 acres in 1944, but has declined steadily since, particularly in East Anglia and in southern counties. Thus East Anglia's share fell from 30% in 1939 to 23% in 1970, while that of Yorkshire rose from 9% to 22%. For, like turnips and swedes, the crop has tended to lose ground to leys, cereals and brassicas, especially kale, on account of its high labour requirements, or to cash root crops such as sugar-beet.

Fig. 83. Bare fallow per 1,000 acres of crops and grass.

Fig. 84. Bare fallow per 1,000 acres of tillage.

Bare Fallow

With the increasing use of artificial fertilizers, greater mechanization and developments in plant breeding, which have made possible a wider range of crops on different soils, the acreage under bare fallow is now much smaller than formerly, although it tends to fluctuate greatly from year to year depending on weather conditions. Chemical sprays for weeds have also made bare fallow less necessary. There were 232,310 acres of bare fallow in 1970, or 1·0% of the crops and grass acreage and 2·3% of the tillage acreage; in 1938 there had been 351,827 acres, or 1·4% and 5·2% respectively, while in 1947, after the severe winter and subsequent flooding, there were 496,504 acres (2·0% and 4·9% respectively). Bare fallow is chiefly important on heavy land or on thin soils where it is difficult to grow root crops (Fig. 83), the highest proportion being in District 102 (5·1%). No clear pattern emerges when bare fallow is shown as a proportion of land under tillage, though if the high values in the uplands (which represent only small acreages) are disregarded, there is a broad correspondence with areas of heavy soil (Fig. 84).

Other Fodder Crops

Kale for Stockfeeding

The most important brassica among other fodder crops is kale, which is especially suitable as winter feed for dairy cattle, although it is also used for other cattle and sheep. The acreage recorded in 1970 was 142,443 acres, or 0·6% of the crops and grass acreage and 1·6% of the tillage acreage. The principal areas for this crop are south and south-west England, although it is also relatively important in Norfolk, east Yorkshire and the Midlands; within the lowlands of southern England, low values are chiefly associated with horticulture (Fig. 85). The highest value recorded is 3% in District 270. This distribution is partly related to the location of dairy farming and partly to climate. Kale is particularly satisfactory in milder areas where the hardier varieties can be left in the ground all winter; it is also fairly drought resistant. The crop is relatively most important in south-west Wales and south-west England, and also in the Pennines where there is only a small acreage under tillage (Fig. 86); the highest value is 50% in District 45.

Cabbage, Savoys and Kohl Rabi for Stockfeeding

Until 1960, these crops were recorded with kale. In 1970, they occupied 7,528 acres, or 0·03% of the acreage under crops and grass and 0·7% of the tillage acreage. These crops are also most important in south-west England (Fig. 87); the highest value recorded is 0·7% in District 299. In relative terms, too, the south-west is the principal area (Fig. 88); the largest percentage of tillage recorded is 2·4% in District 297.

These crops, together with kale, accounted for 99,038 acres in 1938, and the acreage devoted to them increased steadily until about 1960. Kale became popular during the Second World War, for it is high yielding and less costly to produce than root crops, and expansion was particularly marked in south-west England. The acreage has since fallen to less than a third of its former peak.

Beans for Stockfeeding

Beans for stockfeeding are another fodder crop which was formerly much more important, playing a major role in the heavy land rotation of wheat, beans and fallow. In 1970 there were 188,941 acres under beans, or 0·8% of the crops and grass acreage and 1·9% of the tillage acreage, compared with a 1938 acreage of 129,837 acres (0·5% and 1·9% respectively) and a wartime peak of 276,211 acres (1·1% and 2·4%) in 1944. Since the mid-19th century, beans have lost ground to other fodder crops on account of low yields, liability to disease and high production costs (though part of this loss may be attributed to a rise in the acreage returned under mixed corn); but

Fig. 85. Kale for stock feed per 1,000 acres of crops and grass.

Fig. 86. Kale for stock feed per 100 acres of tillage.

Fig. 87. Cabbage for stock feed per 10,000 acres of crops and grass.

Fig. 88. Cabbage for stock feed per 1,000 acres of tillage.

they have recently revived in popularity as a break crop and the acreage trebled between 1963 and 1970. The major areas for beans for stockfeeding lie in an arc from Essex to Holderness (excluding Norfolk) where most districts have 10 or more acres per 1,000 acres of crops and grass, the highest proportion being 86 per 1,000 in District 204 (Fig. 89). Many of these districts have large acreages of heavy to medium, often calcareous, soils; the crop has a strong rooting system and does best on heavy soils rich in lime, and is mainly grown where rainfall is less than 30 inches. The distribution is primarily a southern one, probably because in northern districts winter beans are liable to suffer from frost. The distribution of districts with 20 or more acres

Fig. 89. Beans for stock feed per 1,000 acres of crops and grass.

Fig. 90. Beans for stock feed per 1,000 acres of tillage.

per 1,000 acres under tillage is also very similar, the highest proportion being 10% in District 204 (Fig. 90). There has been little change in the distribution of this crop, with East Anglia providing about two-fifths of the acreage grown in both 1938 and 1970.

Rape for Stockfeeding

Rape is a brassica, primarily of value for sheep feed. It is also widely grown as a catch crop and as a pioneer crop in land improvement, so that some of the acreage sown probably escapes enumeration. The area recorded in 1970 was 46,181 acres, or 0·2% of the crops and grass average and 0·5% of the tillage acreage. Despite the extensive area under rough grazing in the principality, the largest acreage is found in Wales where most upland districts have 1% or more of the crops and grass acreage under

Fig. 91. Rape for stock feed per 1,000 acres of crops and grass.

Fig. 92. Rape for stock feed per 1,000 acres of tillage.

Fig. 93. Rape for oilseed per 1,000 acres of crops and grass.

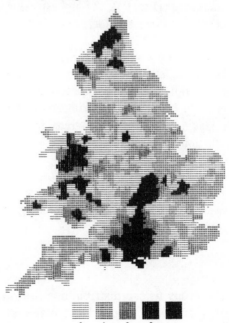

Fig. 94. Rape for oilseed per 1,000 acres of tillage.

107

rape; Northumberland and Cornwall are also important areas (Fig. 91). Rape can be grown under a wide range of conditions, although it is liable to mildew in the south, and its distribution is related chiefly to its use as sheep feed and as a pioneer crop. The highest proportion of the tillage acreage under rape is also to be found in districts in upland Wales, where a fifth or more is under the crop, the highest proportion being 63% in District 346A (Fig. 92).

Rape Grown for Oilseed

Rape grown for oilseed is now separately enumerated; 9,894 acres were recorded in 1970 or 0·04% of the acreage under crop and grass and 0·1% of the tillage acreage. This crop is grown mainly in southern England, especially on the chalk downland of Berkshire and Hampshire; the highest value is 0·9% in District 249 (Fig. 93). The map of percentage of tillage is similar (Fig. 94), except that high values are also recorded (possibly erroneously) in Wales.

Rape was formerly treated as one item. In 1938, 54,401 acres were recorded and a wartime peak of 156,150 acres, or 1·4% of tillage, compared with 56,075 in 1970. Wales's share has increased from 24% in 1939 to 41% in 1970.

Mustard for Seed, Fodder or Ploughing in

The acreage of mustard totalled 15,656 acres in 1970. The crop is grown mainly in and around the Fenland, the western margins of East Anglia and on the Hampshire chalklands (Fig. 95); the highest value is 1·0% in District 156. The map of proportion of tillage is very similar.

Other Crops for Stockfeeding

For completeness, separate maps are included showing the distribution of all other crops not separately enumerated. Those intended for stockfeeding totalled 24,404 acres in 1970. Their distribution is inevitably patchy, but the map of the percentage of tillage locates the chief areas in the uplands (Fig. 96); the highest value is 7% of tillage in District 80.

Other Crops not for Stockfeeding

These crops are likely to be even more heterogeneous and are mainly concentrated in eastern and southern England (Fig. 97): the acreage recorded in 1970 was 17,657 acres.

Crop Combinations

The crops whose distributions are displayed in the preceding maps never occur in monoculture in England and Wales, but compete with other crops. Crops are generally most prominent, therefore, in areas where they have the greatest comparative advantage. Moreover, crops are grown in rotations which are necessary to keep the

Fig. 95. Mustard per 1,000 acres of tillage.

Fig. 96. Other crops for stock feed per 1,000 acres of tillage.

Fig. 97. Other crops not for stock feed per 10,000 acres of crops and grass.

Fig. 98. Number of crops in crop combinations.

109

land in good heart, to prevent the build-up of crop pests and diseases and to utilize the labour force efficiently. It is important, therefore, to have some measure of the way in which different crops are associated in various parts of the country. Figures 99–102 and the tables in Appendix III are an attempt to do this. The crops used are those considered in this chapter, together with those vegetables grown primarily on a field scale, viz., Brussels sprouts, brassicas, carrots, peas for harvesting dry, and green peas, broad beans and French beans for processing. The method used, which is described in more detail in Appendix I, is to rank the crops in descending order of importance according to the acreage they occupy, and then to determine, by the method of least squares, what theoretical combination they most resemble. Areas with similar cropping can then be grouped together. The absence of a crop from a combination does not mean that it is not grown in that district, for small acreages of most crops are grown nearly everywhere, but merely that the crop is not sufficiently important to be included in the combination. Figure 98 shows the number of crops occurring in the identified crop combination in each district and is, in that sense, a measure of the diversity of cropping. It shows a broad three-fold division of England and Wales, with a central belt of very simple crop combinations, extending from north-east England to the south coast, and flanked by belts of much more diversified cropping, the western one embracing the uplands and adjacent lowlands and generally characterized by small acreages, and the eastern including the main areas of cash-crop production, where large acreages of the component crops are grown.

Figures 99–102, on the other hand, show, in descending rank order, the component crops; a common symbol, a dot, is used to indicate those districts where no crop is included in the combination at that level and obviously becomes more frequent in the maps of successively lower ranking crops. Not surprisingly in view of the very large increase in the acreage devoted to barley in the post-war period, it is the leading crop nearly everywhere (Fig. 99). The chief exceptions are a few districts, mainly in the Fenland, where wheat is the leading crop, others in the uplands where oats or fodder crops are more important, a few in the south-west where mixed corn outranks any other crop, and some districts, mainly in eastern England, where field vegetables are the first-ranking crop. In the map of second-ranking crops, wheat replaces barley over most of the central belt, while field vegetables, potatoes and sugar beet are prominent in combinations in eastern England; over most of the western belt, on the other hand, oats, mixed corn and other fodder crops are second crops (Fig. 100). The map of third-ranking crop has no crop recorded in most of the central belt, field vegetables or sugar-beet in eastern England, and oats, mixed corn and other fodder crops in Wales and western England (Fig. 101). With the fourth-ranking crop (Fig. 102), the same pattern is repeated, though the area signifying no crop has greatly extended. The broad pattern is thus one of a central belt, which includes both areas where crops are of only moderate importance and those where a high proportion of farmland is devoted to tillage crops (notably the southern chalklands), where cropping is dominated by barley and wheat; an eastern belt of much more complex crop combinations, where cash

110

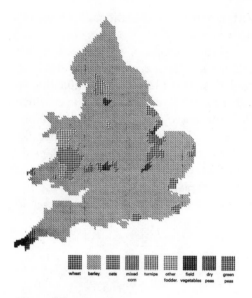

Fig. 99. Leading crop in crop combinations.

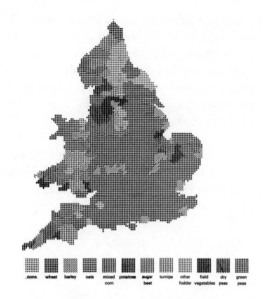

Fig. 100. Second-ranking crop in crop combinations.

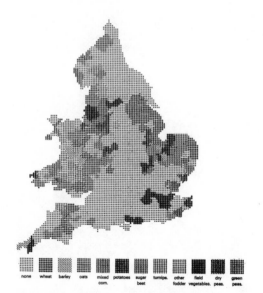

Fig. 101. Third-ranking crop in crop combinations.

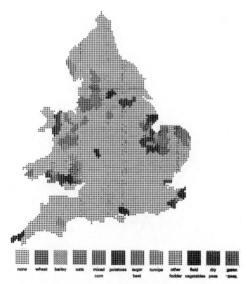

Fig. 102. Fourth-ranking crop in crop combinations.

crops predominate; and a western belt, also with complex combinations of crops, but with mainly small acreages of a variety of fodder crops.

Of course, the patterns could be varied by grouping crops in various ways. For example, all field-scale vegetables could be included as a single item, as could fodder roots, or even all fodder crops; but there would then be little logical justification in including the separate cereals, which dominate nearly all combinations and which cannot, on the information available, be divided into those sold and those consumed on the farm. It was thought best, therefore, to retain the individual crops. Since the procedure does not take account of the acreage devoted to rotation grass, the maps do not show rotations, though it can be inferred that the crops shown do play an important part in rotations. The maps also provide a measure of the relative suitability of different areas for the various crops, as determined by their competitive power in June 1970.

CHAPTER V

Grassland

Grass is the most important crop grown in England and Wales and the one to which most of the country is best suited on climatic grounds; measured in terms of starch equivalent, it is also the cheapest feed for grazing livestock. In this chapter the distribution and importance of the different kinds of grassland are examined. Lucerne, which, unlike clover, is separately enumerated in the census, is also considered.

Grass provides grazing for sheep, cattle and the few remaining horses. Before the Second World War, a large, but unknown, acreage was used as little more than space on which dairy cattle were fed on purchased feed, but this practice is much less common now. Nevertheless, around large towns there is a considerable acreage of poor, neglected grassland in semi-agricultural use, either as accommodation land or awaiting urban development. About a third of the grassland also provides conserved feed, as hay, dried grass or silage, while a smaller proportion is used for seed production.

Although grass accounts for 62% of the acreage under crops and grass, it receives only 30% of the fertilizer applied to the land, and of this twice as much is applied per acre to temporary grass as to the much larger acreage of permanent grass. It is, therefore, not surprising that a survey in the 1950s should have rated only one-fifth of the grassland in England as good and over half as inferior. Grass is probably grown under a wider range of conditions than any other crop and shows a great range of quality; it varies from productive leys and first-class fattening pastures, which will fatten a bullock and a sheep to the acre, to poor permanent grassland on the margins of the rough grazings with only a quarter of the carrying capacity.

Grass comprises both temporary and permanent grass. In theory the distinction between them is clear. Temporary grass is that grown in rotation: it is part of the arable land, to be ploughed up after 1, 2, 3 or more years, succeeded by a variable number of tillage crops and then by a new sowing of temporary grass. Permanent grass is that not in rotation. Yet the dividing line is difficult to draw in practice, partly because of the great regional differences in the length of time for which leys are left unploughed and partly because of the desirability of reseeding 'permanent grass' in western counties at intervals, either directly or after 1 or 2 years of tillage, to prevent further deterioration of quality on the acid soils. There is, in fact, a continuum between 1-year leys on the one hand and grass which is never ploughed on the other, and the exact point at which the line dividing temporary and permanent grass is drawn will

113

depend on the definitions used. Thus, an increase in the acreage of temporary grass and a decrease in that of permanent grass in 1959 are partly attributable to the re-introduction into the census of a category of temporary grass which had been sown 7 years or more, which some occupiers had presumably previously regarded as permanent grass. Nevertheless, while the two kinds of grassland are statistically distinct and while there are important differences between them, they do have many features in common. In view of these statistical ambiguities, it will be useful first to consider them together.

In 1970 there were 13,385,897 acres under grass (excluding lucerne). Figure 103 shows the percentage of the crops and grass acreage under grass throughout the country; this map is, of course, almost the mirror image of the map of tillage (Fig. 48). With a few exceptions, there is a gradient of increasing proportions of grassland from east to west, the highest percentages being found in and around the uplands in Wales, the Pennines and north-west England. In much of the southern Pennines more than 90% of the crops and grass acreage is under grass, the proportion exceeding 99% in District 80. Among the lowland districts, Carmarthen, the Vales of Gwent, south-west Dorset and the Somerset Levels have 85% or more under grass. The only anomalous areas with less than 70% under grass lie in south-west Lancashire, the lower Exe valley, western Cornwall and the rain-shadow of the Welsh uplands. The proportion of grassland is lowest in the Fenland, reaching 5% in District 156, while the only areas between the Thames and the Humber with 25% or more are the Lincoln marshes, Broadland and south Essex. The south-east is also atypical, for, with the exception of Kent, districts in south-east England have 50% or more of their crops and grass acreage under grass.

The map of the 1940 Grassland Survey, published by the Ordnance Survey, shows those areas in which better and poorer quality grassland is to be found. It will be discussed more fully in the section on permanent grass, to which it more properly belongs; but it serves as a reminder, in the absence of any satisfactory alternative, that high proportions of land under grass do not necessarily mean grass of the highest quality.

The predominance of grass in some areas and its minor role in others can be explained partly in terms of suitability for grass farming and partly in terms of unsuitability for other forms of agriculture. The ideal grass climate is one which is not too wet, i.e., not greatly exceeding 40 inches, and not too dry, i.e., not less than 25–30 inches, with fairly frequent and well distributed rainfall and a long season of grass growth. These conditions will, of course, be modified by the availability of soil moisture; thus heavy textured soils and high water-tables offset many of the disadvantages of low rainfall. In the light of these considerations, the country can broadly be divided into three: eastern and south-east England, which are liable to drought in summer and where there are extensive areas of light to medium soils; a middle belt from Northumberland to Dorset where, given suitable soils, climatic conditions are favourable to the growth of good quality grass, particularly in the western part; and the hilly western districts, where high rainfall makes for soil acidity

Fig. 103. Grass per 100 acres of crops and grass.

and waterlogging, which inhibit the growth of the better grasses, but where, at least in the south-west, the growing season is long, so that grass is abundant if not of the highest quality. Most of the anomalous districts in the west are to be explained either by climatic advantages, e.g., somewhat lower rainfall, as in the Welsh borderland, or by lighter soils; those in the east with above average proportions of grass are primarily due to the presence of poor heavy land, as in the Weald, or to a high water-table, as in Romney Marsh. The east is thus not well suited to grass but is generally suited to tillage, grass is generally preferred on the heavier soils of midland and southern

Fig. 104. Temporary grass per 100 acres of grass.

Fig. 105. Grass for mowing per 100 acres of grass.

England, while western areas, although not ideal for good quality grass, are even less suitable for tillage.

In 1970 26% of the grassland in England and Wales was temporary grass, but the proportion which is regarded as being in the rotation varies considerably throughout the country (Fig. 104). The highest proportions are found chiefly in eastern arable areas, on the chalklands of southern England and in the south-west peninsula, in all of which there are extensive areas where temporary grass accounts for 40% or more of all grassland. In the Hampshire chalkland the proportion rises to over 50%, and the highest percentage is found in District 263 (59%). The lowest percentages are found in those districts where most of the crops and grass acreage is under permanent grassland, viz., the Pennines, parts of Wales, Somerset and eastern Dorset.

There is no longer any clear-cut distinction between pasture, which is only grazed, and meadow, which is only mown, for most grassland is grazed and is also mown at intervals; 33% of the grass acreage was mown in 1970 and, while there are few districts in which less than 30% of the grass is mown, there are regional differences (Fig. 105). Conditions for hay or silage-making are generally unsatisfactory in western districts, particularly at higher elevations, because of difficulties of harvesting and the poor quality of much of the grass. In eastern areas much of the small acreage of grass is in access fields, or is required for summer grazing and is consequently not available for mowing. Similarly, only a small proportion of the grassland in the fattening districts of the east Midlands is mown. The highest proportions of mowing grass are in the south, particularly in Dorset, Gloucester and Wiltshire, where 35% or more of the grass acreage is mown, the highest values being 48% in District 262. In the Pennines, hay plays an important part in farming systems, but in most northern districts the proportion cut for hay is lower.

While most of the acreage mown in 1970 was cropped for hay, there are no published data on the regional importance of hay and silate-making, although the proportion cut for hay in Wales was a little higher than that in England. While silage-making has been increasing in popularity in the post-war period, the number of farmers making silage and the amount made are still small. Sample data from the Milk Marketing Board suggest that silage-making is most common in midland and southern England. In eastern England there are problems of the availability of labour, for silage-making tends to clash with other urgent tasks.

The areas under grassland changed considerably over the previous three decades. In 1938 there were 17,702,891 acres (excluding lucerne), or 72% of the crops and grass acreage, of which temporary grass accounted for 11% and mown grass for 30%. There was a considerable reduction during the Second World War to a minimum acreage in 1944 of 12,726,709 acres (including lucerne), or 52%, 23% and 33% respectively. Thereafter, the grassland acreage rose steadily, although an increasing proportion was temporary grass. Thus, while the acreage under grass in 1970 was only a quarter less than that in 1938, the proportion of temporary grass was two and a half times greater. The proportion of the total grass acreage in western and northern counties was generally higher in 1970 than in 1939, while that in midland and southern counties was lower.

Permanent Grass

Much of the land in permanent grassland will probably never be regularly ploughed, chiefly on account of intractable and poorly drained soils, the hilly and dissected nature of the terrain, or because the land is low-lying and liable to flooding. Some fields are permanent grassland because they are access fields or accommodation land, or because they are far from the farmstead. But large acreages can be ploughed, as the wartime plough-up campaign showed.

There were 9,885,465 acres of permanent grass in 1970, or 42% of the crops and

Fig. 106. Permanent grass per 100 acres of crops and grass.

20   35   50   65   80

grass acreage. Figure 106 shows the proportion of permanent grass in the various districts. The highest values are found in and around the industrial Pennines, where in several districts 90% or more of the land in crops and grass is under permanent grass, but the uplands generally stand out as areas with high proportions of permanent grass (although the acreages may be small because of the predominance of rough grazing). Lowland areas where permanent grass is particularly important include South Wales and the Somerset Levels. The lowest proportions are in the Fenland, where several districts have less than 5%, for the land here is too valuable to use as grassland. Areas in the west with less than 50% of the crops and grass acreage in permanent grass are the south-west peninsula, south-west Lancashire and the Eden valley. In south-central England, the chalklands stand out, with a third or less under permanent grass; by contrast, this proportion is exceeded in most of south-east England and in the east Midlands, where there are large stretches of clay soils.

No adequate measure is available of regional differences in the productivity of permanent grassland. The Grassland Survey map provides an indication of areas where good grassland was to be found in 1940, although it is not a map of the actual distribution of grassland. Ryegrass, which will not survive long in acid soils, can be used as an index of good quality grass, and agrostis or bent, which can tolerate much more acid soils and wetter conditions, as an index of poor quality grassland. Good grassland was found mainly in Cheshire and Lancashire, Somerset, the east Midlands and Northumberland; medium-quality grassland was characteristic of most other parts of Lowland Britain, except the south-east; and poor-quality grassland and rough grazing were typical of the west and north.

Hay yields also provide an indication of the productivity of grassland, although regional differences in the length of the growing season and in the extent to which land can be grazed in wet weather without poaching, and the varying demands of livestock do not make these yields a wholly adequate measure of the amount of grass available. Thus, while grass growth normally ends in October in north England, it may continue into December in the south-west.

The 5-year average yield in England and Wales (1966–7 to 1970–1) was 28·9 cwts. per acre, compared with 32·6 cwts. in Northumberland. With the exception of Holland and Cambridgeshire, in both of which the acreage under meadow hay is small, high yields are obtained in the belt of counties from Cheshire to Somerset and in the north, rather than in those counties where the proportion of permanent grassland is highest.

High percentages of permanent grass occur around the uplands because arable farming is difficult there, but conditions are far from ideal for grass growth, with acid soils, high rainfall and humidity and a short growing season. In the Pennines, and to a lesser extent in South Wales, there is the additional handicap of atmospheric pollution which increases acidity, checks grass growth and adversely affects clover; on an experimental farm on Rossendale it was estimated that 5 cwts. of soot fell per acre each year. But permanent grass is abundant in the Pennines, not only for climatic reasons, but also because it is an area of small, often part-time farms; for such small farms frequently have high proportions of permanent grass. In eastern England, above average proportions of permanent grass are associated either with alluvium, where the water-table is high, as in Broadland or Romney Marsh, or with poor heavy land, as in the Weald. The most extensive areas of good permanent grassland occur in areas of moderate rainfall and on those soils with a good water-retaining capacity which are naturally fairly well drained. Although some of the differences between western districts may be due to differences in the interpretation of what is permanent grass, the low proportions which mark off the south-west peninsula from other western areas are

Fig. 107. Permanent grass per 100 acres of grass.

Fig. 108. Permanent grass for mowing per 100 acres of permanent grass.

due chiefly to the important place that temporary grass occupies in the farm economy there.

There were 2,576,201 acres of permanent grass mown in 1970, or 26% of the total. Figure 108 shows that the distribution of districts with different proportions of mowing grass bears a strong resemblance to the corresponding map of all grass (Fig. 105), although the range of differences between districts is considerably greater. Proportions are highest in South Wales and southern England particularly in east Somerset, Wiltshire and Dorset where 45% or more of the permanent grass is mown, and around the Pennines. In southern districts the longer growing season and the quality of the grass give better prospects of a good hay crop or silage; in the Pennines, although conditions are not generally favourable for haymaking, tradition plays an important part and the dales are dotted with isolated hay barns. In the south-west the high proportion of grass under temporary grass may explain the low percentage of permanent grass mown; for the permanent grass acreage contains a high proportion of poor grassland which cannot be mown, of which some is little better than rough grazing. The proportion of mowing grass is also low in areas of summer fattening on grass and in eastern districts where there is little grass. A noticeable feature of the map is the marked difference between the two major dairying areas centred on Cheshire and Somerset respectively, for in the former less than a quarter of the permanent grass is mown and in the latter, more than two-fifths.

In 1938 there were 15,823,862 acres of permanent grass, or 64% of the crops and grass acreage, while 27% was mown. Since then the permanent grass acreage has fluctuated more markedly than the acreage of all grass. The acreage fell rapidly during the Second World War, reaching a minimum of 9,755,731 acres, or 40% of the crops and grass acreage, in 1944. Extensive areas of heavy land in the Midlands, together with stretches of downland and much land further west, were ploughed up, so that the proportion of permanent grass in western and midland counties decreased by as much as 30%. After the war the acreage of permanent grass rose fairly steadily as land was once more laid to grass; this change was most marked in Wales and the north-west and least in eastern counties. In the late 1950s and early 1960s, however, this fall was reversed in eastern and midland counties, though it continued further west. As a result, western and northern counties now have a larger share of the permanent grass acreage than in 1938, and the acreage in some eastern counties is now smaller than it was at the peak of the wartime ploughing campaign.

Temporary Grass

Temporary grass is usually more productive than permanent grass, but its establishment involves greater expenditure in ploughing and seeding, loss of production, and a greater risk. One of the most striking changes in the use of agricultural land in recent years has been the great increase in the acreage under temporary grass, which more than doubled between 1938 and 1958. In part this increase is illusory, resulting from changes in the interpretation of what should be returned as temporary grass; but there is no doubt that a large part of the increase is real, reflecting the more widespread adoption of ley farming. In ley farming, all the ploughable land is ploughed, and temporary grass leys alternate with tillage crops. This contrasts with the system in which there is a sharp distinction between permanent grassland, which is rarely, if ever, ploughed, and arable land (although 1-year clover leys may be grown as part of the arable rotation). There is still considerable scope for increasing the acreage under leys at the expense of permanent grass.

In 1970 there were 3,500,432 acres of temporary grass (excluding lucerne), or 15% of the crops and grass acreage and 25% of the arable acreage. The distribution of temporary grass is somewhat different from that of permanent grass, although for both crops there is a contrast between low proportions in the east of the country and high

Fig. 109. Temporary grass per 100 acres of crops and grass.

proportions elsewhere (Fig. 109). Some of the differences have obvious explanations, for the percentage of land under temporary grass must be low in those districts, such as the Pennines and east Somerset, where permanent grass accounts for 80% or more of the acreage under crops and grass. The most important area for temporary grass is the south-west peninsula, where it accounts for more than 30% of the acreage under crops and grass. Other major areas are the chalklands of southern England; parts of central and south-east Wales; north Cheshire; the Lake District and north-east England. The Fenland, on the other hand, has less than 5% of the crops or grass acreage under temporary grass, a proportion which falls to under 1% in some districts.

The 5-year average yield of 'seeds' hay (i.e., hay from temporary grass) was 34·9 cwts. per acre, and the high proportion of temporary grass mown probably makes these yields a more reliable index of productivity than those of meadow hay. Apart from Holland and east Sussex, highest yields are obtained from northern England; yields are also above average in eastern England. The highest county figure was 41·5 cwts. in the East Riding.

There are many reasons for these regional differences in the importance of temporary grass. Most of the major areas have moderate rainfall and fairly tractable soils. In the south-west there has long been a tradition of ploughing grassland; on the chalklands higher rainfall makes it possible to grow grass more easily than on light soils further east, while improvements in grasses and in the techniques of ley farming have made it easier to establish good grass; the high proportions in the north also reflect a long-established practice of ley farming. Low proportions in eastern England are generally associated with areas of light to medium soils on which cash cropping is more profitable than leys, although a higher proportion than formerly has been made necessary in some districts by considerations of crop hygiene, e.g., the incidence of eel-worm in the Fenland.

When expressed as a percentage of the arable acreage, temporary grass is seen to be relatively more important in the west than appears from Figure 109. Even in the Pennines and the Somerset Levels, where the acreages of both arable and temporary grass are small, more than 60% of the arable is in temporary grass, and similar proportions are found over most of Wales, north-west and south-east England (Fig. 110). The highest proportion, 87% in District 80, is somewhat misleading because of the very small acreage involved. Temporary grass occupies only a third or less of the arable land on the chalklands, despite the large acreage grown. In East Anglia and the Fenland, both the acreage grown and the proportion of the arable are small, the lowest values being found in Fenland districts.

A total of 1,877,203 acres, or 54% of the temporary grass acreage, were mown in 1970 to make hay, silage or other forms of preserved grass. The distribution of districts with high proportions of temporary grass for mowing is rather patchy, partly because small acreages receive the same weighting as large (Fig. 111). High proportions are characteristic of much of northern England and the north Midlands, almost the reverse of the permanent grass map (Fig. 108); the highest value is 100% in

123

District 43. Percentages are below average in the south-west and in Wales, where conditions are not ideal for haymaking; they are also low in east and south-east England where there are relatively few grazing stock.

Traditionally, clover and rotation grass were sown in eastern England in 1-year leys as part of the Norfolk four-course and other similar rotations, while long leys and the periodic reseeding of grass were largely confined to western counties. Figure 112 shows that, with the exception of a few areas such as south-west Lancashire, where 1-year leys play an important part in farming peatland, and the chalk downs, 1-year leys are still characteristic of eastern areas, although a small acreage is grown in nearly

Fig. 110. *Temporary grass per 100 acres of arable land.*

Fig. 111. *Temporary grass for mowing per 100 acres of temporary grass.*

every district. The principal areas extend from the Yorkshire Wolds, through Lincolnshire, to Hertfordshire, and then to Suffolk, and have 25% or more of their temporary grass acreage in 1-year leys. In the Fenland and Norfolk, on the other hand, proportions are comparatively small, for part of the acreage formerly used for clover is now devoted to peas for processing.

A map of the proportion of longer leys is the mirror image of that of 1-year leys, and a map of leys sown in 1963 or earlier shows this contrast in more extreme form (Fig. 113), with values of 25% and over in Wales and in south-west and north-west England. The degree to which such older leys can properly be regarded as temporary grass must, of course, depend on farmers' intentions.

There were 1,901,303 acres of temporary grass (excluding lucerne) in 1938, or 8%

of the crops and grass acreage and 21% of the arable, while 62% was mown. During the early war years the acreage decreased to a minimum of 1,767,462 acres, or 15% of the arable, but the exhaustive cropping of these years made it necessary to lay an increasing acreage of land down to grass, so that by 1944 there were 2,970,978 acres (including an unknown, but probably small, acreage of lucerne which was not then separately distinguished) or 12% of the crops and grass acreage and 20% of the arable. After the war the temporary grass acreage continued to rise fairly steadily until the early 1960s. Though it has fallen since then, as more land has been ploughed for cropping, it is clear that ley farming is now far more widely established than it was in

Fig. 112. One-year leys per 100 acres of temporary grass.

Fig. 113. Grass sown in 1963 or earlier per 100 acres of temporary grass.

1938. This increase may be partly explained by improvements in varieties of grass and in techniques of grassland husbandry, by farmers' experiences of ley farming during the war years, by the influence of advocates such as Sir George Stapledon, and by the stimulus of subsidy payments for ploughing up grassland which has been down more than 3 years. Such subsidies were abandoned in 1967 because it was felt they had served their purpose in encouraging farmers to adopt ley farming. The temporary grass acreage is now more evenly distributed throughout the country than it was before the war, for there has been both a relative and an absolute increase in the importance of temporary grass in midland counties. In 1970 both eastern and western counties contributed a smaller proportion of the total acreage (though a larger acreage) than they did in 1939; thus Devon and Cornwall accounted for only 14% in

1958, compared with 19% in 1939, and Norfolk, Suffolk and Essex for 4% compared with 10%.

Lucerne

Lucerne is strictly a tillage crop, but it is perennial, being usually left down for 4 to 7 years and it is often sown with grass in a ley, fulfilling a similar role to clover in fixing atmospheric nitrogen, so that it seems more appropriate to consider it under the heading of grassland. There were 37,371 acres of lucerne in 1970, or 0·2% of the crops

Fig. 114. Lucerne per 1,000 acres of crops and grass.

Fig. 115. Lucerne per 1,000 acres of arable land.

and grass acreage, 0·3% of the arable acreage and 2·3% of the acreage under temporary grass. It is not possible to say what proportion of the lucerne acreage is accounted for by pure crops of lucerne and what proportion by mixtures with grasses. The principal areas for lucerne are in eastern England (Fig. 114). The largest acreages are grown in west Norfolk and Suffolk, the highest proportion of the crops and grass acreage being 5% in District 178. There are also areas with 0·5% or more in upland Wales, though the actual acreages recorded are small. Most of the large acreages occur in areas of light soil, often underlain by chalk; because of its deep rooting habit, lucerne is particularly suitable for areas of porous soils, where the water table lies some distances below the surface, as in the Breckland. The ecological preferences of lucerne are for a fairly dry and sunny climate and neutral or alkaline soils, though these can hardly be reconciled with the high values recorded for Wales and north-west England in Figure 115, which were not a feature of the map for 1958.

Although a larger acreage is grown now than before the war, it is not possible to follow changes in the acreage and distribution of lucerne throughout this period, since it was included in the temporary grass acreage between 1943 and 1949. There were 31,274 acres in 1938, or 0·1% of the crops and grass acreage, 0·4% of the arable acreage and 1·6% of the temporary grass acreage. During the 1950s there was little change, but the acreage in 1970 was only a third of that in 1958. East Anglia remains the principal area, though its share has declined from 42% in 1939 to 35% in 1970, a trend contrary to the general tendency for a declining acreage to be concentrated in the area best suited to the crop.

This summary of the main features of the distribution of grassland is much shorter, in relation to the length of discussions on tillage and horticultural crops, than the importance of the subject warrants. Its brevity has been dictated in part by the small number of studies in which the reasons for these regional differences in grassland husbandry have been investigated, but chiefly by the nature of the census data available.

CHAPTER VI

Horticulture

Although they differ considerably from each other, horticultural crops merit separate treatment because their distribution is more localized, in terms both of area and of the number of holdings on which they are grown, than that of most farm crops. They also require large inputs of labour and often capital, and they are grown for direct human consumption. Those grown in the open tend to vary greatly in yield from year to year as a result of differences in weather conditions, to which many of them are highly sensitive. Since they are also generally high yielding, production is usually a better index than acreage; unfortunately, no regional production figures are published, nor is there any indication how yields vary from place to place. Although hops are generally regarded as farm crops, they share many of the characteristics of horticultural crops, and it will be convenient to consider them here.

These crops vary greatly in their requirements and in the conditions under which they are grown. Some outdoor vegetables are produced on large arable farms for about £70 per acre whereas the more careful husbandry of specialized growers lifts expenditure to about £120 per acre; marketing costs are additional. By contrast, the growing of out-of-season flower crops indoors can cost up to £20,000 per acre. Outdoor vegetables generally need little specialized equipment, but provision of environmental control in modern glasshouses entails an inclusive expenditure of not less than £25,000 per acre, while modern orchards require an investment of £400 per acre (excluding land) before they bear profitable crops. There is also a wide variety of growers, differing in degree of specialization, acreage of holding and size of business. The specialist fruit grower supplies 60% of the fruit grown, but the arable farmer grows 80% of the coarse vegetables, such as cabbage, savoys and Brussels sprouts, while market gardeners, who concentrate on choicer vegetables, grow 50% of other vegetables.

In 1970 there were 729,761 acres under horticultural crops, or 3% of the crops and grass acreage. Of this total, 67% were under vegetables, 21% under orchards, 4% under small fruit (of which part is also included in the orchard acreage), 2% under hops, 3% under flowers grown in the open, 2% under hardy nursery stock and less than 1% in crops in glasshouses.

While it is not surprising that a small acreage of horticultural crops should be grown in nearly every district in view of both the distribution of the population and the wide range of crops included under horticulture, two major areas stand out: the Fenland and north Kent (Fig. 116). Other important areas are: east Norfolk, east

Fig. 116. Horticultural crops per 1,000 acres of crops and grass.

Lindsey, the Humber warplands, Bedfordshire, Worcestershire, south-east Essex, Middlesex and south Buckinghamshire. There are also outlying areas of lesser importance in south-east Lancashire, Herefordshire, south-east Hampshire and west Cornwall. Apart from the uplands, the areas with little horticulture are in the lowlands of the west and north, the southern chalklands and the clay lowlands of the Midlands. Many small pockets of horticulture, such as that at Coombe Martin in north Devon, cannot be shown at this scale.

With the exception of hops, all the main branches of horticulture are represented to some extent in the principal horticultural districts, but their relative importance varies considerably between regions. The eastern region makes the biggest contribution to horticultural output, with the production of vegetables as the leading activity. In the south-east, which is the most important region measured in output per acre, fruit accounts for more than half the output, as it does also in the west Midlands. In the south-west, output is fairly evenly divided among fruit, vegetables, flowers and glasshouse crops, much of the orchard acreage being of little value.

Many of the reasons for this distribution will emerge in the subsequent analysis. The Kent, Worcester and Middlesex areas are all long established and pre-date the coming of the railways; but the scale of activity has increased greatly since the end of the 19th century, except in Middlesex, where the spread of London has taken much land formerly used for horticulture. The character of horticulture has also changed; thus in Middlesex and Worcestershire the proportion of the horticultural acreage devoted to vegetables has greatly expanded. Horticultural areas, such as south-west Lancashire, have risen to prominence with the growth of the great industrial cities of the north. In some places, e.g., mid-Bedfordshire, where there is evidence of market gardening in the 17th century, the seeds of horticultural specialization were sown early, but the areas became important only in the second half of the 19th century; in others, e.g. the Fenland, an interest in horticulture developed during the great agricultural depression in arable farming at the end of the 19th century. Although the acreage of horticultural crops in these areas has increased and their relative importance has changed, no major concentration of horticulture has been established since the 1930s.

In 1938 there were 588,811 acres under horticultural crops, or 2·4% of the acreage under crops and grass; of this, vegetables accounted for 44% and orchards for 43% (although the acreage under vegetables is not strictly comparable with that in 1970). The acreage rose rapidly during and immediately after the Second World War, reaching a maximum of 900,676 acres in 1948, of which 62% were vegetables; at this time horticultural crops were more widely grown, especially in the main arable areas. Since 1948 the acreage has declined and there has been a marked eastward movement in the centre of gravity of horticultural cropping compared with 1939; thus East Anglia and the Fenland accounted for 22% of the acreage in 1939 and 37% of the much larger acreage in 1970. This easterly shift is, in fact, much more important than these figures suggest, for the share of gross output contributed by these areas is proportionately higher than their share of the horticultural acreage.

130

Vegetables and Flowers Grown in the Open

Vegetables and flowers account for most of the area under horticultural crops, although the proportion and composition vary considerably throughout the country. They comprise a variety of crops with widely different physical and economic requirements. For purposes of this analysis, flowers and bulbs grown in the open have been included with vegetables, since they share many of the same cropping characteristics and occupy a similar place in rotations. Among the vegetables there is a broad two-fold division into coarse vegetables, mainly grown on a field scale by arable farmers, and the choicer vegetables, which require more attention and are often grown by specialist market gardeners. Early crops of vegetables may also be grown by market gardeners, although most of the main crop may be produced by arable farmers. There are also areas which specialize in early or off-season production, as in west Cornwall. Market gardeners tend to grow a wider range of crops and the acreages of their holdings are often small; arable farmers generally grow only a few crops which occupy only a part of their holdings, and the early districts also tend to concentrate on a few crops.

The distinction between ordinary farm crops and coarse vegetables and between the latter and market gardening crops is not easy to draw in practice; for vegetables may be grown by farmers as a speculative venture and ploughed in or fed to stock if prices are not favourable, and the range of crops grown by farmers has steadily been extending and now includes vegetables such as celery and carrots which were formerly grown principally by market gardeners. As a result, the range of crops grown on market gardens has tended to decrease.

There were 487,375 acres recorded under vegetables in June 1970, or 2·0% of the crops and grass acreage and 4·8% of the tillage acreage. This figure is somewhat misleading, for not only is there a certain amount of double cropping, but some spring vegetables have been cleared and some winter crops not yet planted at the time of the June census. The cropped acreage, i.e., the total area cropped with vegetables in the course of a crop year, was some 523,000 acres in 1970. The difference between the June and the cropped acreages is most marked in the case of spring cabbage, winter cauliflower, lettuce, turnips and swedes.

Some vegetables are grown in nearly every district in England and Wales, but the chief areas of intensive, specialized production of vegetables are the Fenland, mid-Bedfordshire, the Evesham district, the Humber warplands and east Lindsey; other important areas with 5% of their crops and grass acreage under vegetables are the West Riding, south-west Lancashire, Norfolk, south Essex, east Kent and south Hampshire (Fig. 117). Some vegetables are grown around all the major cities, and there are isolated pockets, such as west Cornwall. The only major areas of arable farming where vegetables are unimportant are the chalklands of southern England.

The map showing the proportion of the tillage acreage occupied by vegetables is

Fig. 117. Vegetables per 1,000 acres of crops and grass.

5   25   50   100

very similar (Fig. 118), for values are highest in the principal vegetable growing areas already noted, in all of which 10% or more of the acreage under tillage is devoted to vegetables, while in most of East Anglia, Kent and Lindsey, the proportion is 5% or over; the highest value is 37% in District 154. On this basis of comparison, several other areas are more prominent, especially those on the fringes of large cities; but a large proportion of vegetables grown in East Anglia, Lindsey and the West Riding are for processing (Fig. 119).

Most of the major centres of vegetable growing have a long association with horticulture, although west Cornwall is almost entirely a product of the railway age.

Fig. 118. Vegetables per 1,000 acres of tillage.

Fig. 119. Vegetables for processing per 100 acres of all vegetables.

Vegetable growing around London, although long-established, has been gradually displaced into south Buckinghamshire, south-west Essex and north-west Kent as the metropolis has expanded. Vegetable growing on a field scale is of more recent origin. The depression in arable farming in the late 19th century led farmers to grow as vegetables either crops such as cabbage which they had formerly grown only for stockfeed, or new crops such as sprouts which also required relatively little attention; similarly, experience gained in growing sugar-beet from the 1920s onwards encouraged farmers to adopt tap-rooted vegetables such as carrots and parsnips. The range of vegetable crops grown by farmers has extended, particularly through the establishment of canneries and later quick-freezing plant in the main arable areas (Figs. 45 and 119); such crops are often grown on contract, a practice which removes the possibilities of large profits, but ensures a market at an agreed price. Many of these field crops, such as peas for canning, fit well into arable rotations. As a result of this expansion of vegetables growing on arable farms, the eastern counties' share of the vegetable acreage has grown both relatively and absolutely. Moreover, most of the established market gardening areas are surrounded by a zone where farmers have adopted a limited range of vegetables; thus the growing of Brussels sprouts has spread from mid-Bedfordshire into north Hertfordshire.

It is difficult to find convincing explanations of the location of many areas of vegetable growing. The various crops have a wide range of soil requirements, and vegetables are to be found on soils running the gamut of light sands to peats and heavy loams; but, in general, most vegetables are grown on soils of light to medium texture which are free draining and can be easily worked at all times of the year. Thus, around Birmingham and the Black Country the light soils of Lichfield and Bromsgrove are important. In many of the old-established areas soils have been considerably modified; in parts of the rhubarb growing areas south-east of Leeds as much as 6 inches of screened ash have been added to the soil to improve its drainage. Mid-Bedfordshire received large quantities of manure from London during the 19th century, but the cessation of this supply with the disappearance of cart horses and town dairies has presented serious problems of maintaining soil fertility; shoddy is now used in large quantities as an organic manure.

Climatic controls, particularly wind and humidity, are generally less rigorous than in the case of fruit growing. Most vegetables are favoured by the climatic conditions which suit arable crops in general, and, while frost and drought are more serious hazards in vegetable production than in arable farming, the high value of the crops can often justify measures to offset climatic hazards. Thus irrigation water is widely used to increase yields, and irrigation equipment is now common on market gardens. The milder winters of coastal areas, particularly in the south-west, make them important for early production; but early vegetables are also grown in inland districts.

Location is an important consideration in the growing of crops such as lettuce, whose freshness affects their market value, for crops grown for canning or freezing, which must be processed quickly, and, to a lesser extent, for bulky, low value crops, e.g. cabbage, which will not bear the cost of transport over long distances. Yet many bulky vegetables, such as cauliflower and carrots, travel long distances; for example, there is regular movement from the Fenland to Scotland. Remoteness certainly restricts the range of vegetables which can be profitably grown in the south-west peninsula and makes it difficult for isolated though climatically suitable areas to compete. Cornwall is the only county which depends primarily on rail transport to its markets, and it seems probable that for many vegetables links with wholesalers are more important than actual location, although proximity to Covent Garden is said to confer advantages by the ability to meet sudden demands for extra supplies of a crop.

Although the harvesting of some crops, such as vining peas, has been mechanized, vegetables generally make heavy demands on hand labour. All the main vegetable growing areas are therefore distinguished by relatively high densities of both casual and permanent labour (Figs. 28 and 29). The holdings of many vegetable growers, particularly in the Fenland, Bedfordshire and Evesham, are small, and this characteristic has been accentuated by the creation of statutory smallholdings. The payment of piece-work in vegetable production and opportunities for part-time and casual employment for women and high earnings from piece-work do much to offset the pull of urban wage rates which might otherwise attract labour.

Over the previous 32 years the acreage under vegetables varied greatly. In 1938 there were 261,974 acres or 1·1% of the crops and grass acreage and 3·9% of that under tillage; there was a great expansion in vegetable production during and after the war, particularly that of coarse vegetables, and a maximum June acreage of 560,484 acres, or 2·3% of the crops and grass acreage, was recorded in 1948. Since then there has been a considerable reduction in acreage, although it remains well above pre-war figures. Furthermore, owing to rising yields and to the changing composition of the vegetable acreage, production of many crops has either risen or fallen less steeply. There has also been a shift towards eastern arable counties; Norfolk accounted for 6·1% of the total acreage of vegetables in 1939 and 14·9% in 1970, while the percentage in Lindsey rose from 5·6% to 12·5%.

Brussels Sprouts
Brussels sprouts were grown on 53,645 acres in 1970, or 11·0% of the vegetable acreage. The crop is found on both arable farms and market gardens. It can be grown on a wide range of soils provided they are well drained and have a satisfactory lime content. Small acreages are grown throughout the lowland, but production is highly localized in two areas, the one centred on Biggleswade, the other on Evesham; in the former, six districts accounted for 39% of the total acreage, and in the latter, four districts contain 19% of the total (Fig. 120). While there are no obvious reasons for this concentration, Bedfordshire has long been pre-eminent, and the crop has now spread into neighbouring counties, particularly the arable fringe of north Hertfordshire, partly to avoid club root, partly because it fitted into arable rotations, and partly because teams of skilled pickers were available. The growing of Brussels sprouts has similarly spread from the Vale of Evesham on to the Cotswolds, which provide somewhat later crops and so extend the period of marketing. An increasing acreage (19% of the 1970 crop) is grown for processing, especially in East Anglia and Lindsey (Fig. 121).

There were 37,965 acres grown in 1939, or 15·3% of the vegetable acreage; the total rose to a maximum of 54,748 acres in 1948. Apart from the rise of East Anglia and Lindsey, the relative importance of the different areas has changed little.

Cabbage and Savoys for Human Consumption
A total of 42,949 acres of cabbage and savoys for human consumption was recorded in June 1970, or 18·8% of the vegetable acreage. These figures minimize the

importance of the crops, because different varieties are grown at different seasons, so that the cropped acreage of some 61,100 acres is considerably larger. Cabbages and savoys can be grown on a wide range of soils, but, as with many vegetables, they do best on loams which are not too acid. Figure 122 shows the acreage recorded in June 1970 under these crops; they are prominent in most of the horticultural areas and are least important in upland districts of the north-west and on the clay lowlands.

Figure 123 shows the distribution of summer and autumn cabbage and savoys, as recorded in the June census. The crops are grown mainly by arable farmers and are

Fig. 120. Brussels sprouts as a percentage of Brussels sprouts in England and Wales.

Fig. 121. Brussels sprouts for processing as a percentage of Brussels sprouts for processing in England and Wales.

important around the Wash and in districts surrounding the main centres of population, for they are bulky crops of low value and cannot bear heavy transport charges. Winter cabbages and savoys, as recorded in the June census, are also grown around the conurbations, but they are prominent in coastal districts, particularly Lincolnshire, where mainly savoys are grown, and Norfolk, where cabbages account for most of the acreage. Spring cabbage is grown chiefly in coastal districts of Kent (15% of the June acreage) and Holland (20%); the cropped acreage in 1969/70 was 25,700 acres, compared with an acreage of 6,877 of remaining spring cabbage recorded in June.

In June 1939 there were 44,062 acres of cabbage, savoys, kale and sprouting broccoli recorded, or 16·9% of the vegetable acreage; the maximum acreage during

Fig. 122. Cabbage and savoys as a percentage of cabbage and savoys in England and Wales.

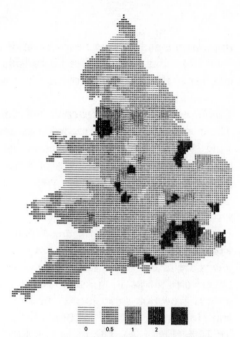

Fig. 123. Summer and autumn cabbage and savoys as a percentage of summer and autumn cabbage and savoys in England and Wales.

Fig. 124. Summer and autumn cauliflower as a percentage of summer and autumn cauliflower in England and Wales.

Fig. 125. Winter cauliflower as a percentage of winter cauliflower in England and Wales.

the wartime and post-war expansion of coarse vegetable growing was 77,672 acres in June 1947, but there has been a considerable decline since. There have been no major changes in the distribution of these crops.

Cauliflower, Sprouting Broccoli and Kale

There were 32,138 acres recorded as under cauliflower, sprouting broccoli and kale in June 1970, or 6·6% of the vegetable acreage; but this figure, too, is misleading, since much of the winter cauliflower (or broccoli, as it is generally known) is not then in the ground. The cropped acreage was some 39,500 acres, of which summer and autumn cauliflower accounted for 49% and winter cauliflower for 45%. Summer cauliflower is grown in most of the main vegetable areas, especially the silts of the Fenland, where Holland alone accounts for more than two-fifths of the acreage, north Kent and south-west Lancashire; much of this production is market oriented, while in Holland, high yields are also a factor (Fig. 124). The map of winter cauliflower as recorded in June understates the position, but shows the three principal areas, west Cornwall, where the crop is grown on small farms, the Isle of Thanet, where production is on large farms, and Holland; Cornwall accounts for 25% of the acreage, Kent for 17% and Holland for 14% (Fig. 125). The main considerations in this distribution are the avoidance of severe frosts and the early onset of the growing season; but the Cornish area did not rise to prominence until after the coming of the railway, while the crop became important in Thanet only after it had been successfully introduced by an individual in 1870.

There were 18,924 acres recorded under cauliflower in the June census of 1939, or 7·6% of the vegetable acreage; this acreage increased to a maximum of 34,438 acres in 1948 (with a cropped acreage of 54,800 acres), fell, and then began to rise again in 1953. These minor changes hide a marked increase in the relative importance of Holland, whose share of the total acreage has risen from 5% in 1939 to 28% in 1970.

Carrots

In 1970 there were 37,610 acres under carrots, or 7·7% of the acreage under vegetables. The crop is grown in four principal areas: west Norfolk and north-west Suffolk; the peat Fens; the Humber warplands; and south Lancashire (Fig. 126). There is also an important area in the Suffolk Sandlings. Early carrots are also grown chiefly in these districts. The crop is mainly grown on sandy soil, peat or alluvium; deep, light soils are preferred, for on heavy soils there is a danger of roots splitting and the crop is harder to clean. Supplies of water for washing are also an important consideration. In Norfolk much of the crop is grown on a large scale, and an increasing acreage (31% in 1970) is grown for processing, especially in Norfolk, which has 52% of the acreage for processing.

There were 16,134 acres of carrots in 1939, or 6·2% of the acreage under vegetables. The acreage under carrots rose to a maximum of 39,857 acres in 1946 and has since fluctuated around 30,000 acres. The major change in distribution has been

138

the rise of western Norfolk as the principal area of production; Norfolk's share of the acreage under carrots rose from 16% in 1939 to 44% in 1970, while that of the East Riding (mainly the Vale of York), which is one of the older established areas, fell from 15% to 2%.

Green Peas for Market

Peas picked green for market accounted for 12,968 acres in 1970, or 2·7% of the acreage under vegetables. This crop is grown mainly near the principal towns and cities, particularly in the West Riding, which has 28% of the acreage, and Essex, which has 13%; the Evesham District in Worcestershire is also important (Fig. 127). Early peas are a speciality in the Bridgwater area. Green peas do best on deep loamy soils, but the essential considerations are supplies of labour for picking the crop and quick access to markets to preserve its freshness.

There were 60,587 acres of peas picked green for market in 1939, or 23·2% of the vegetable acreage. Since 1945, when there were 63,090 acres, there has been a steady decline in the acreage grown, owing to the large increase in the acreage of peas grown for canning or quick-freezing. Between 1939 and 1970 the share of Essex and the West Riding, now the principal producers, rose from 36% to 41%, though the acreage grown fell by 75%.

Green Peas for Processing

There were 127,537 acres of peas for processing in 1970, or 26·2% of the acreage under vegetables; an increasing proportion of peas for processing (56% in 1965) is quick-frozen. The crop is primarily grown on fairly large farms in the main arable areas of eastern England, the counties of Holland, Lindsey, Norfolk and Suffolk accounting for 65% of the acreage (Fig. 128). There are outlying areas of production in the West Riding (8%) and in Kent (2%); only small acreages of peas are grown elsewhere. Most of the crop is grown under contract, and the location of canning and freezing plant is an important control. It is grown on a variety of soils, for there are obvious advantages in spreading the harvest. Production is highly mechanized, so that labour resources are not a major factor.

The acreage of peas for canning or quick-freezing has risen fairly steadily since it was first recorded as 18,414 acres in 1941, or 5·7% of the acreage under vegetables. Norfolk's share of the acreage rose from 15% in 1941 to 28% in 1970, and Lindsey's from 16% to 29%.

Peas for Harvesting Dry

Like green peas for market, peas for harvesting dry are a crop of declining importance. In 1970 there were 73,298 acres, or 15·0% of the acreage under vegetables. The crop is almost entirely confined to eastern arable districts, especially in those counties between the Thames and the Humber (Fig. 129). Climate is the principal control over this distribution, since the crop requires rather similar conditions to those needed for

Fig. 126. Carrots as a percentage of carrots in England and Wales.

Fig. 127. Green peas for market as a percentage of green peas for market in England and Wales.

Fig. 128. Green peas for processing as a percentage of green peas for processing in England and Wales.

Fig. 129. Peas for harvesting dry as a percentage of peas for harvesting dry in England and Wales.

wheat, with good harvesting conditions in August. The crop also does best on fairly heavy soils, as in Holderness. It is widely grown throughout the arable districts, but the principal areas are the Fenland, which accounts for almost a fifth of the acreage, Lindsey, east Essex and Holderness.

There were 71,156 acres of peas for harvesting dry in 1941, when the crop was first separately enumerated; during and after the Second World War, production was stimulated by guaranteed prices and the acreage rose rapidly to a maximum of 180,171 acres in 1949, or 34% of the vegetable acreage. Since then it has declined steadily; so, too, has the share of Kesteven and Lindsey, which accounted for 44% of the acreage grown in 1939 and 10% of that grown in 1970, while the proportion grown in Holland and Norfolk increased from 23% to 32%, and that in Essex from 3% to 15%.

Parsnips
There were 6,600 acres under parsnips in 1970, or 1·4% of the acreage under vegetables. The crop is grown both in the principal areas of intensive market gardening in the southern half of the country and also, like carrots, on the peats of the Fenland and on the light soils of west Norfolk and north-west Suffolk, where it is similarly grown by large growers (Fig. 130).

There were 6,485 acres under parsnips in 1941, when the crop was first separately recorded, or 2·6% of the acreage under vegetables. The acreage rose to a maximum of 8,876 acres in 1948, but remained fairly constant from 1951 until the late 1960s, when it began to rise again. Over these three decades, Norfolk's share of the acreage grown rose from 23% to 42%.

Beetroot
Beetroot totalled 7,679 acres in 1970, or 1·6% of the acreage under vegetables. Beet is widely grown, but, unlike other root vegetables, it is rarely a farm crop because it needs careful handling (Fig. 131). The principal areas are therefore the market gardening districts around London, mid-Bedfordshire, and the Humber warplands, especially the Isle of Axholme, south-west Lancashire and the Fenland. About 44% of the crop is grown for processing, mainly in Cambridgeshire, the Isle of Ely, Lindsey and the West Riding, which account for 63% of that acreage.

There were 8,579 acres of beet grown in 1941 when the crop was first separately recorded, or 2·7% of the vegetable acreage. The acreage under beet rose to a maximum of 13,738 acres in 1946 and has since fluctuated around 8,000 acres.

Turnips and Swedes for Human Consumption
There were 6,508 acres (or 1·3% of the acreage under vegetables) recorded under turnips and swedes for human consumption in June 1970; in 1962, when they were separately enumerated, swedes accounted for two-thirds of the acreage. Much of the acreage sown escapes enumeration in the June census; the cropped acreage of turnips

and swedes was 9,700 acres. These crops are widely grown, both in eastern arable counties and near large towns, notably in south-west Lancashire (Fig. 132); but the principal producer is Devon, which is also an important producer of turnips and swedes for fodder and which accounts for nearly a quarter of the June acreage. The red soils of Devon are particularly suitable for swedes, and the Devonshire crop commands a market premium because of its quality.

The acreage recorded under turnips and swedes in June 1941 was 9,007 acres, or 2·8% of the acreage under vegetables. It increased to a maximum of 12,724 acres in 1947, but has since declined, fluctuating around 5,000 acres. The importance of Devon has steadily increased for, while the county accounted for only 3% of the June acreage in 1941, the proportion had risen to 23% in 1970.

French Beans

There were 17,652 acres of French beans in June 1970, or 3·6% of the acreage under vegetables. This crop is now grown almost entirely for processing, only 6% being grown for market. Production is therefore mainly concentrated on large arable farms within range of processing factories, notably in west Norfolk, and also in south Yorkshire, Bedfordshire, south Essex and West Sussex (Fig. 133).

French beans were not separately enumerated before 1953, being returned with runner beans. When these two crops were first enumerated in 1941, their combined acreage was 8,931 acres, or 20% of the acreage under vegetables. The acreage rose to 12,815 acres in 1948 and, after a decline, began to rise again in 1953, when there were 1,978 acres under French beans. Norfolk's share has risen from 10% in 1953 to 32% in 1970.

Broad Beans

There were 11,317 acres under broad beans in 1970, or 2·3% of the acreage under vegetables. Broad beans are grown both for market and for processing, and exhibit two rather different distributions, as do many crops grown for both outlets. Twenty-seven per cent of the acreage grown in 1970 was under broad beans for market, much of it in the principal market gardening areas in the Fens, mid-Bedfordshire, the Evesham district and around London; early production is characteristic of Somersetshire (Fig. 134). The 73% of the crop grown for processing was located mainly in the Fenland, south-west Lancashire, the Evesham district and south Kent (Fig. 135). Broad beans are grown on a wide range of soils, but do best on calcareous loams and marine silts. The influence of the location of the processing factories is clear from Figure 135; labour is also important in the harvesting of this crop, and the availability of gang labour is also a factor.

Runner Beans

There were 8,007 acres under runner beans in 1970, or 1·6% of the acreage under vegetables. This crop, too, is grown both for processing and for market, but, unlike

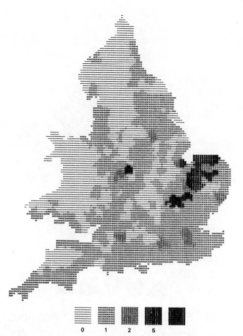

Fig. 130. Parsnips as a percentage of
parsnips in England and Wales.

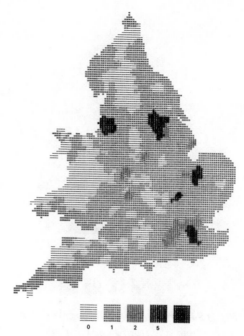

Fig. 131. Beetroot as a percentage of
beetroot in England and Wales.

Fig. 132. Turnips and swedes as a
percentage of turnips and swedes in
England and Wales.

Fig. 133. French beans as a
percentage of French beans in
England and Wales.

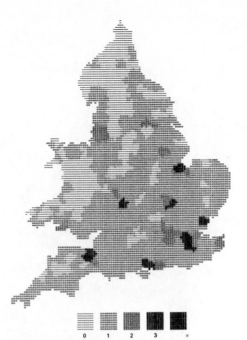

Fig. 134. Broad beans for market as a percentage of broad beans for market in England and Wales.

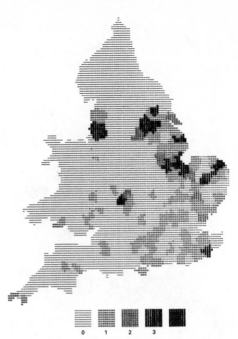

Fig. 135. Broad beans for processing as a percentage of broad beans for processing in England and Wales.

Fig. 136. Runner beans for market as a percentage of runner beans for market in England and Wales.

Fig. 137. Runner beans for processing as a percentage of runner beans for processing in England and Wales.

broad beans, only 18% of the crop is intended for processing. Runner beans for market are grown mainly in the market gardening areas, notably in the Vale of Evesham, mid-Bedfordshire, south Essex and Kent (Fig. 136). Runner beans for processing, on the other hand, are grown mainly in south Yorkshire, the silt Fens and east Norfolk (Fig. 137).

There were 2,996 acres under runner beans in 1953 when the crop was first enumerated separately; the acreage has declined from a peak of 10,174 acres in 1959. Essex's share has fallen from 24% in 1953 to 12% in 1970, while that of Norfolk and Suffolk has risen from 6% to 13%, reflecting the increasing importance of processing.

Salad Onions

There were 2,329 acres of onions grown for salad in June 1970, or 0·5% of the acreage under vegetables; the cropped acreage, at some 3,900 acres, is considerably larger. Salad onions are grown principally around the major urban centres, particularly London, and in the areas of specialized market gardening, notably the Vale of Evesham (where District 102 alone accounts for 21% of the June acreage), mid-Bedfordshire and the Fenland (Fig. 138).

Like beans, onions have not been separately recorded under different heads in the published statistics. In 1939 there were 1,719 acres under salad and bulb onions, or 0·7% of the vegetable acreage. During the Second World War there was a considerable expansion to a maximum of 15,262 acres in 1944, or 3·5% of the vegetable acreage, resulting from a rapid rise in the acreage of bulb onions. Since 1948 the acreage has fallen fairly steadily. Only 3% of the acreage under onions was grown in the Fenland in 1939, but much of the wartime acreage was planted in the Fenland, which accounted for 36% of the total in 1944 and 43% in 1947; since then the proportion has fallen to 23%.

Onions for Harvesting Dry

The were 3,084 acres of bulb onions for harvesting dry in 1970, or 0·8% of the acreage under vegetables. They are grown principally in the Fenland, where Holland accounts for 29% of the acreage and Cambridgeshire and the Isle of Ely for 19% (Fig. 139). The crop is best suited to medium loams and does not do well on light sands or heavy clays. About 90% of consumption is imported, as weather at harvest time is generally not sufficiently warm or dry.

Celery

There were 4,558 acres of celery recorded in June 1970, or 0·9% of the acreage under vegetables; this figure is rather less than the cropped acreage of some 5,000 acres. Production of this crop is also highly localized, the chief area being around Littleport; 62% of the June acreage is grown in three districts on the peats of the Fenland (Fig. 140). The crop is also grown on the mossland of south Lancashire and on the silts of the Fenland and of the Humber warplands. Associated with the field production of the crop on the peats of the Fenland is a highly specialized glasshouse industry, centred on

145

Fig. 138. Onions for salad as a percentage of onions for salad in England and Wales.

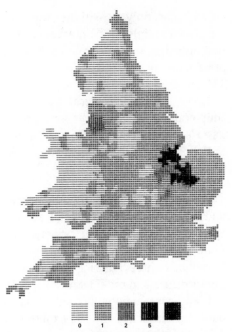

Fig. 139. Onions for harvesting dry as a percentage of onions for harvesting dry in England and Wales.

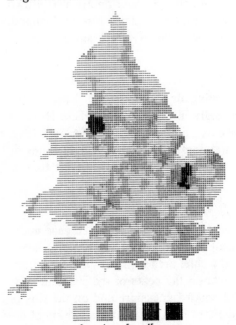

Fig. 140. Celery as a percentage of celery in England and Wales.

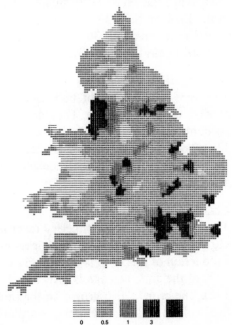

Fig. 141. Lettuce as a percentage of lettuce in England and Wales.

Whittlesey, which produces young plants for transplanting. Deep organic soils with a fairly high water-table provide an excellent medium for the growth of celery, while on the silts conditions favour the growing of self-blanching celery on a large scale. The existence of canneries in the Fenland is a further reason for localization of production of celery in this area.

In June 1939, 6,654 acres of celery, or 2·7% of the vegetable acreage, were recorded, with a maximum of 7,264 acres in 1946. The southern Fenland has long been the centre of this industry and its relative importance has changed little.

Lettuce (Not under Glass)

In June 1970 there were 12,183 acres recorded under lettuce or 2·5% of the vegetable acreage. This figure greatly understates the total acreage, for the cropped acreage is some 19,200 acres, crops of lettuce being grown throughout the year. Production of the crop is widely spread, some being grown around most towns, but the principal districts lie around the conurbations, particularly Greater London and south Lancashire, and, to a lesser extent, in other major market gardening areas (Fig. 141). It can be grown on most soils, provided these are not too acid or dry. Proximity to markets is an important consideration, and only north-east England and South Wales cannot supply a large part of their requirements.

There were 5,873 acres recorded in June 1939, or 2·4% of the acreage under vegetables. The acreage increased to a maximum of 8,758 acres in 1946, declined to 6,674 in 1955 and then rose in the late 1960s to a maximum in 1970; the share of the Home Counties has declined from over half the acreage in 1939 to just over a quarter in 1970.

Composition of the Vegetable Acreage

The composition of the acreage under vegetables in the major growing areas thus varies considerably. In Worcestershire, for example, the principal crops are Brussels sprouts, green peas for market, cabbages and runner beans (in that order); in the Isle of Ely, dry peas, carrots and celery; and in Holland, dry peas, cauliflower, peas for canning or quick-freezing, cabbages and broad beans. The proportion of the vegetable acreage occupied by the principal vegetables also varies considerably. In Bedfordshire, Brussels sprouts alone account for 55% of the vegetable acreage, the remainder being occupied largely by cabbages (8%), peas for harvesting dry (5%), green peas for processing (5%), lettuce (3%), and runner beans (3%); in Kent, on the other hand, no crop is so important, for cauliflowers account for 17%, cabbages for 14%, peas for harvesting dry 12%, Brussels sprouts 10%, green peas for processing 10%, runner beans 6%, and lettuce 3%. There are also considerable differences in the extent to which highly localized crops dominate the cropping of the major vegetable-growing areas; thus, while 18% of the acreage under green peas for processing is in Norfolk, where they account for 32% of the acreage under vegetables in the county, the 48% of

147

the celery in England which is grown in Cambridgeshire and the Isle of Ely accounts for only 8% of the vegetable acreage in that county.

Hardy Nursery Stock
A total of 15,880 acres were returned under hardy nursery stock in 1970. Nurseries are widely distributed throughout the lowlands, the most important area being in and around London, particularly on the sandy soils of west Surrey where there are many nurseries, though the silt Fens, south Hampshire and West Sussex are also important (Fig. 142). There were 10,532 acres in 1939, and the extent of land under nursery stock was also reduced during the Second World War to a minimum of 4,927 acres in 1944. Since then it has risen steadily. The leading position of Surrey has been maintained throughout.

Bulbs Grown in the Open
In June 1970 there were 15,027 acres of bulbs grown in the open. Sixty per cent of this acreage was in Holland on the silts of the Fenland, 20% being in District 155, and 16% in the adjacent counties of Cambridge and Norfolk; a further 6% was grown in Cornwall (Fig. 143). Bulb growing was established in south Lincolnshire in the late 19th century, and the silts provided near-ideal conditions, without waterlogging in winter and facilitating lifting in summer. Many of the bulbs are sold to florists in the same area, who force them in heated glasshouses in winter. Mechanization of cultivation and the need for long rotations has led to the adoption of bulb growing on large farms in the area. In Holland 64% of the acreage in September was under daffodils and 32% under tulips. The emphasis in west Cornwall, with its small farms, is on the production of early daffodils for sale as flowers, and sandy loams are favoured; 76% of the September acreage is under daffodils, and much of the remainder is under irises and anemones.

There were 7,656 acres under bulbs in 1939. The acreage was curtailed in the interests of food production during the Second World War and reached a minimum of 1,846 acres in 1944; it has risen steadily since the prominence of the Fenland has become more marked; thus Holland accounted for 47% of the acreage under bulbs in 1939 and 60% in 1970.

Other Flowers Grown in the Open
In June 1970 there were 4,183 acres of other flowers not under glass. They are widely grown, particularly in south-east England, where London provides the largest and wealthiest market in the country (Fig. 145); south Lancashire also serves a large urban market. Other important areas include the Fenland and Cornwall, where the emphasis is on early flowers. There were 5,815 acres of other flowers in June 1939; the acreage was reduced to 1,345 acres in 1943. After surpassing the post-war acreage, the area devoted to other flowers is declining as growers change to more profitable nursery stock.

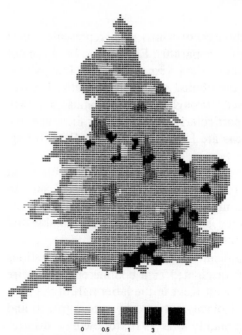

Fig. 142. Nursery stock as a percentage of nursery stock in England and Wales.

Fig. 143. Bulbs as a percentage of bulbs in England and Wales.

Fig. 144. Glasshouses as a percentage of glasshouses in England and Wales.

Fig. 145. Other flowers as a percentage of other flowers in England and Wales.

Glasshouses

In June 1970 there were 4,470 acres of glasshouses, of which 96% represented fixed glasshouses, three-fifths of them served by heating apparatus. Figures for 1939 are not strictly comparable, but 3,268 acres of crops under glass were recorded in June of that year. Cultivation under glass is both widely distributed and highly localized; small acreages of glasshouses occur in most districts throughout the lowlands, but two-thirds are in east and south-east England, particularly in the Lea valley and the Worthing area (Fig. 144). Other important areas are Lancashire and the East Riding, and the remaining market gardening areas.

Glass provides an opportunity for cropping throughout the year, particularly in heated glasshouses, and several different crops may be grown at different seasons. In July 1970 52% of the glasshouse acreage was under tomatoes, 9% under cucumbers and 6% under chrysanthemums. In winter, lettuce is the most important vegetable and a larger acreage is under flowers and foliage plants than under vegetables.

As with much horticultural cropping, it is difficult to find satisfactory explanations of the distribution of glasshouse cultivation. It originated as a commercial enterprise in the Worthing area, the Lea valley and north-west Kent in the latter half of the 19th century. Around London it has been steadily displaced outward by urban growth and by high land values. The Blackpool area arose mainly after 1920, following a disaster in local strawberry stocks, while the growth of cultivation under glass in the Hull area seems to have been due primarily to Dutch immigrants in the 1930s. Proximity of markets is still important, possibly because of the increase in direct marketing, while for some crops, e.g., tomatoes, local supplies command a substantial premium over produce coming from further afield. Individual initiative and location of markets do seem to have played a major role in the development of areas of specialized cultivation under glass.

In view of the high investment per acre, insulation from climatic conditions and the use of sterilized soil, it might appear that physical factors are of comparatively little importance. Nevertheless, climatic conditions, particularly temperature, wind and light intensity, are of considerable importance. Heating costs are a major element in variable costs and are lowest along the south coast, which also has high sunshine values. The Lea valley suffers serious loss of sunshine from atmospheric pollution, and the acreage under glasshouses is declining. Other centres of glasshouse cultivation are similarly far from being ideally located.

Vegetables for Market and for Processing

Attention has been drawn on a number of occasions to the contrast between the location of crops grown for marketing in an unprocessed state and that of crops grown for processing, and Figures 146 and 147 show these differences for vegetable crops as a whole. Vegetables grown for market are grown mainly in the principal market gardening areas, including the Vale of Evesham and mid-Bedfordshire, and other vegetable growing districts in fairly close proximity to the major conurbations, though given the speed and flexibility of modern road transport, such locations are not themselves of any great importance. Vegetables for processing, on the other hand, are grown mainly in east Lincolnshire, east Norfolk and Suffolk (Fig. 147), and location may often be critical, notably in respect of green peas, which must be processed within a short time of picking and are generally grown within 15–20 miles of processing plant.

Processing of vegetables has become of major importance since the Second World War, although both canning and freezing plant existed pre-war. The tonnage of vegetables canned increased tenfold from the mid-1930s to the mid-1960s, though two-thirds of the vegetables used are imported (for beans in tomato sauce and processed

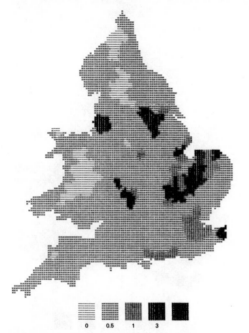

Fig. 146. *Vegetables for market as a percentage of vegetables for market in England and Wales.*

Fig. 147. *Vegetables for processing as a percentage of vegetables for processing in England and Wales.*

peas); the main home-grown vegetables canned are green peas, broad, French and runner beans and carrots. Quick-freezing of vegetables is, to a large extent, complementary to that of fish, and many of the plant are located at east-coast ports (Fig. 45); the principal vegetables quick-frozen are green peas, beans, and Brussels sprouts.

Small Fruit

Small fruit shares characteristics of both vegetables and orchards, for while the plants are perennial, they often occupy a place in arable rotations. There were 33,145 acres under small fruit in 1970, or 0·1% of the acreage in crops and grass, and 0·3% of the tillage acreage. The growing of small fruit is largely confined to southern England (though it is interesting to note that over 90% of the raspberry crop is grown in Scotland). The three main areas are East Anglia and the Fenland; Kent; and the south-west Midlands (Fig. 148), with the highest proportion of District 189 (4%). There are minor concentrations elsewhere and these are often more localized than the map suggests. The distribution of the proportion of the tillage acreage under small fruit (Fig. 149) is very similar to that shown on Figure 148. Only 5% of the acreage is now grown under orchard trees. Forty-eight per cent of the acreage is devoted to strawberries, 31% to blackcurrants, 12% to gooseberries and 4% to raspberries.

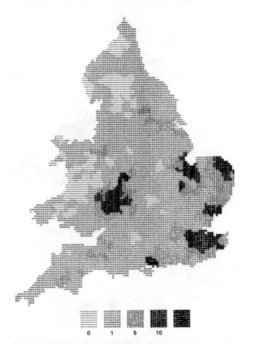

Fig. 148. Small fruit per 1,000 acres of crops and grass.

Fig. 149. Small fruit per 1,000 acres of tillage.

The location of small fruit growing owes much less to tradition than does that of top fruit, and many important areas have risen to prominence only in this century. Physical requirements are generally similar to those of orchards, though small fruit is less susceptible to high rainfall. Much is grown on light to medium soils, as on the silts of the Fenland and the loams of east Norfolk. Small fruit is grown on a variety of holdings; it fits in well with the labour requirements of top fruit, and it is grown by smallholders, by market gardeners and by arable farmers in East Anglia and the west Midlands. Small fruit is either marketed fresh or preserved, either as jam or as canned or quick-frozen fruit, so that both urban markets and the location of processing plant are important, especially for highly perishable fruit such as strawberries.

Horticulture

There have been marked fluctuations in the acreage under small fruit. In 1938 there were 49,302 acres, but during the Second World War the acreage fell to a minimum of 29,855 acres in 1945; it rose again to a peak of 51,170 acres in 1950, and has since fallen, although production is above pre-war levels. The contribution of the principal areas has not changed greatly, although Kent's share has fallen from 25% of the total acreage to 19%.

Strawberries

The 15,761 acres of strawberries grown in 1970 were concentrated in four main areas; the silts of the Fenland; north-east Norfolk; Kent; and south-east Hampshire (Fig. 150). Norfolk, which includes part of the Fenland, has 21% of the acreage, as does Kent. There are smaller concentrations in the Vale of Evesham, the Cheddar area of Somerset, south-west Essex and the Tamar valley, and some strawberries are grown throughout East Anglia and the south-west Midlands. Much of the crop is produced by small growers, but larger acreages are also grown for jamming and canning, which are thought to absorb between a third and a half of the crop.

Strawberries are grown on a wide range of soils; they can tolerate poor drainage, some of the best quality fruit being grown on the clay-with-flints on the North Downs, but they are not suited to thin soils, calcareous soils or shallow clays. A range of sites is advantageous in main crop areas since it extends the short season. Aspect was important in the location of the area of early production in the Tamar valley and Cheddar areas, and freedom from late frosts is obviously desirable. Before the advent of cloches, the south-western area had undeniable advantages in earliness, the Tamar strawberries ripening in the first week of June, a week earlier than those in Hampshire and Cheddar and two weeks earlier than the Kent crop. Much of the crop in the Fenland and East Anglia is processed. The crop requires abundant labour for picking (five to six pickers per acre); but there are many areas where the crop could be grown, and the high degree of localization owes much to advantages arising from being first in the field.

The acreage has fluctuated from 18,732 acres in 1939 to 10,454 acres in 1944, rising to a post-war peak of 21,055 acres in 1950; but a higher proportion is now grown in the Fens and East Anglia. Yields and production are considerably higher than pre-war.

Blackcurrants

There were 10,393 acres of blackcurrants in 1970. This crop is to be found mainly in Kent, the eastern part of East Anglia and in the south-west Midlands (Fig. 151). Blackcurrants are tolerant of poor drainage, but they do best on medium to light soils, as on the loams of east Norfolk. They can also tolerate higher rainfall than top fruit. Although blackcurrants were popular earlier in the century as an undercrop in orchards, most of the acreage is now to be found on ordinary arable farms, where blackcurrants are grown in rotation, having been adopted as a tillage crop during the

153

Fig. 150. Strawberries as a percentage of strawberries in England and Wales.

Fig. 151. Blackcurrants as a percentage of blackcurrants in England and Wales.

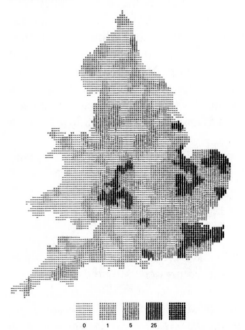

Fig. 152. Gooseberries per 1,000 acres of gooseberries in England and Wales.

Fig. 153. Raspberries as a percentage of raspberries in England and Wales.

great depression in arable farming; on such farms blackcurrants may be the only fruit grown. Seventy-seven per cent of the crop is now grown for processing and the existence of factories, as at Coleford in Gloucestershire, has also played a part in localizing production.

The acreage fell from 10,378 acres in 1939 to 8,397 acres in 1945, rising again to a post-war peak of 16,026 acres in 1950. Kent and Norfolk accounted for 44% of the acreage in 1939 and 36% in 1958 and a higher proportion of the crop is now grown in the south-west Midlands. Yields are considerably higher than pre-war.

Gooseberries

There were 3,989 acres under gooseberries in 1970. The principal centres of production are the silts of the Fenland, north-east Norfolk, Kent and the Vale of Evesham; Norfolk, Cambridgeshire and the Isle of Ely account for 31% of the acreage and Kent for 21% (Fig. 152). In the mid-1960s almost a third of the crop was said to be grown within 10–15 miles of Wisbech, and District 189 had 10% of the crop in 1970. Gooseberries require deep, well-drained soils, such as those derived from Fen silts, to produce good fruit. While gooseberries are no longer required on any scale for jam-making, their production is encouraged in the two main areas by the way it fits in with other labour requirements, particularly among strawberry growers in the Fenland, and by the location of canneries, to which most of the crop is now sent.

There were 9,137 acres of gooseberries in 1939 and the acreage has fallen steadily since, although production is now higher than pre-war. Kent's share of the declining acreage fell from 32% in 1939 to 21% in 1970 while that of the Isle of Ely and Norfolk (largely the Fen silts) rose.

Raspberries

There were 1,248 acres of raspberries grown in 1970. The main centre is north-west Kent, where District 233 had 14% of the crop, but raspberries are also grown in east Norfolk and Suffolk and in the south-west Midlands (Fig. 153). The crop grows best where summers are cool, as in Scotland, and requires deep, well-drained soils with good moisture holding capacity. Raspberries have heavy labour requirements for picking and, since they are highly perishable, there are advantages in locations near urban markets. Much of the crop, particularly on the larger holdings, is grown for processing and this is probably the major outlet in the remoter areas.

There were 4,135 acres under raspberries in 1939, and the crop shows a similar war-time decline to a minimum of 2,044 in 1945 and a post-war peak of 4,348 acres in 1951. The crop has suffered from competition from Scottish growers. Kent's share of the diminished acreage, 33% in 1970, is almost the same as in 1939 (32%).

Orchards

In 1970 there were 153,725 acres of orchards, or 0·7% of the acreage under crops and grass. There are two main orchard areas, Kent and the south-west Midlands

155

Fig. 154. Orchards per 1,000 acres of crops and grass.

(Worcester–Herefordshire) (Fig. 154). Kent has long been the principal county for orchards and contains 33% of the acreage under top fruit; district boundaries are not very suitable for analysing its distribution within the county, but orchards are to be found mainly on the dipslope of the North Downs, on the ragstone ridge and on the High Weald, the chief concentrations being around Faversham–Sittingbourne and Maidstone. Orchards are widespread in Worcestershire, lowland Herefordshire and north-west Gloucestershire, while there are numerous small farm orchards in Somerset and south Devon. Other important, but smaller concentrations are to be found in the Wisbech area, in north Cambridgeshire and in south-east Essex. Little fruit is grown in

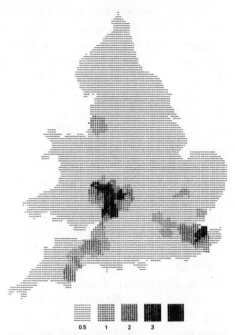

Fig. 155. Commercial orchards as a percentage of commercial orchards in England and Wales.

Fig. 156. Non-commercial orchards as a percentage of non-commercial orchards in England and Wales.

northern England or Wales, or on the chalklands or the clay lowlands, and most of the acreage here is in small farm orchards.

Orchards in England and Wales vary greatly in quality and composition, though these differences have been reduced by the extensive plantings of the post-war period and by the grubbing up of old orchards with the encouragement of government grants. Separate returns are now made of orchards which are grown commercially and those which are not grown commercially; the former accounted for 139,660 acres in 1970, or 91% of the total. Their distribution thus closely resembles that of all orchards (Fig. 155). They are highly localized both by area and by holding; less than 4% of all holdings have commercial orchards. The distribution of non-commercial orchards is more widespread and the largest acreages occur in west and south-west England, especially in the counties of Gloucester, Hereford and Worcester and in Devon and Somerset, though only in north England and Wales are there greater acreages of non-commercial than of commercial orchards (Fig. 156). They are found on less than 3% of holdings and are much less common on larger holdings.

Both physical and historical explanations may be offered for the distribution of orchards. Climatic controls are clearly important, areas with high rainfall and cool summers having little fruit. Most commercial orchards are to be found east of the 30-inch isohyet and in the warmer and sunnier southern half of the country. High rainfall encourages diseases and makes fruit growing highly speculative; sunshine and warmth are necessary for ripening, while spray and wind largely exclude top fruit from exposed coastal areas. Kent is as well suited as any area in south-east England, although it does not appear to enjoy any special climatic advantage, while Worcestershire is nearer the rainfall limit. Most of the fruit-growing areas are liable to experience the late frosts which are primarily responsible for the great fluctuations in yield from year to year, although these are partly due to bad siting. Frost damage can best be avoided by careful selection of sites where there is free air drainage without excessive exposure.

Waterlogging of the soil, even at infrequent intervals, is a serious hazard and orchard trees all do best on deep, well-drained soils; 12 inches has been suggested as a minimum depth of soil. Different crops and different varieties of the same crop vary considerably in their tolerance of poor drainage, cherries being the most sensitive and cooking apples and plums the least. Kent is favoured with an abundance of well-drained soils, particularly those developed on the Thanet Sands, the brick-earths of the North Downs, the ragstone of the Lower Greensand, and the sandstones of the High Weald. The loams of east Essex, the silts of the Fenland, the terrace gravels of the Avon and the sandstones of north-west Worcestershire also give rise to well-drained soils. Conversely, areas which are low-lying and have heavy soils or a high water-table have generally been avoided, as have the shallow soils of the chalk. Nevertheless, it must not be assumed that all orchards are correctly sited; once planted, orchards tend to persist for the life of the trees.

Both Kent and Worcestershire are old-established fruit growing areas, with reputations dating at least from the 17th century. Both have been favoured by their location near major markets, although for Worcestershire this has been true only since the rise of the Birmingham–Black Country conurbation in the 19th century. These urban markets have, in return, provided casual labour for fruit-picking. Both areas have probably benefited from their reputation and from the development of specialized services and skills. In Kent the fruit-growing area has gradually extended eastward as communications have improved and as London has spread outward, and, apart from its legacy from the past, location is probably now of little importance in explaining the present distribution of orchards. The growing of fruit for cider and perry making in the west country was also well established in the 18th century, but fruit growing around

Wisbech began only in the 1850s and did not become important until the 1880s; it is interesting to note that an important figure in this development was a migrant from Kent. Most commercial production in East Anglia is also of recent origin.

The acreage under orchards changes little from year to year. There were 251,330 acres in 1938, or 1% of the acreage under crops and grass; the acreage rose, probably in part as a result of more complete enumeration, to a maximum of 277,655 acres in 1952, but has fallen steadily since, though rising yields have largely compensated for this loss. Both the composition and distribution of the acreage under orchards have changed considerably; many old orchards, particularly cider and perry orchards, have been grubbed up and there has been much planting of new orchards in eastern counties, particularly of dessert apples which give a much higher return than cooking apples. As a result there has been a marked easterly shift in the centre of gravity of orchard fruit production. Kent now has 51% of the acreage under orchards, compared with 28% in 1939, and Essex, Norfolk, Suffolk, Cambridge and the Isle of Ely 19% compared with 13%. By contrast, the share of Devon, Gloucestershire, Somerset and Worcestershire has fallen from 29% to 19%.

Dessert Apples

The acreage of dessert apples grown commercially in 1970 was 57,581 acres, or 41% of the acreage of all commercial orchards; these figures compare with a total of 56,400 acres, or 21% of the total orchard area, in 1947. Kent is by far the most important county, with 42% of the total acreage, followed by Essex, with 13% (Fig. 158). Dessert apples are more susceptible to adverse climatic conditions than cooking apples, particularly to heavy rainfall and little sunshine, and are less tolerant of poor drainage, although texture as such does not seem to matter greatly provided drainage is good. Eastern districts are therefore preferred and there has been considerable planting of new orchards of dessert apples in the post-war period. The acreage of Cox's Orange Pippin, the most important variety, continues to rise, though the total acreage of dessert apples is declining.

Cooking Apples

There were 27,941 acres of cooking apples in 1970, or 20% of the commercial orchards. Kent is also the major producer of culinary apples, accounting for 60% of the acreage in 1970, followed by Norfolk and Essex with 7% each and Worcestershire with 3% (Fig. 159). The acreage has fallen steadily from 72,100 acres in 1947, when these represented 27% of all top fruit, as cooking apples have been replaced by more profitable dessert apples. As with all other orchards, these comparisons of acreages are not strictly valid, since the 1970 figures relate to commercial orchards only and those for 1947 to all orchards.

Cider Apples

The acreage of cider apples in 1970 was 14,130 acres, or 10% of the acreage under

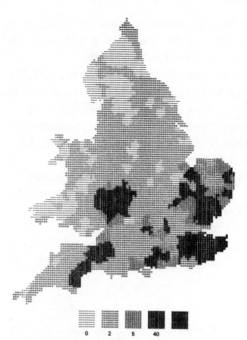

Fig. 157. Apples per 1,000 acres of crops and grass.

Fig. 158. Dessert apples per 1,000 acres of crops and grass.

Fig. 159. Cooking apples per 1,000 acres of crops and grass.

Fig. 160. Cider apples per 1,000 acres of crops and grass.

160

commercial orchards. Herefordshire, with 43% of the acreage, is the leading county, followed by Somerset, with 22%, and Devon, with 16% (Fig. 160). These areas have all been long established, but the acreage has been considerably reduced since the end of the 19th century; the crop was once grown in eastern districts, but seems to have been excluded by competition from culinary and dessert apples. In 1947 there were 51,300 acres, or 20% of the acreage of all top fruit, but there has been a steady decline since.

Dessert and Cooking Pears

In 1970 there were 13,888 acres of dessert and cooking pears, or 10% of the acreage of commercial orchards. Kent is also the leading county for such pears, accounting for 61% of the acreage under pears, followed by Cambridgeshire (6%) and Essex and Worcestershire, with 5% each (Fig. 161). Conditions suitable for pears are very similar to those for apples, although they are regarded as a more difficult crop to grow and are rarely the principal crop of commercial growers. They are, however, a useful complement to apples, since good years for pears seem to coincide with bad years for apples. The acreage in 1947 was 14,800 acres, or 6% of the acreage of all top fruit, and rose steadily during the 1950s, but has since declined.

Perry Pears

There were only 928 acres of perry pears in 1970, or less than 1% of the acreage of commercial orchards. Somerset was the leading county, with 53% of the acreage, while Gloucestershire and Herefordshire had 15% each (Fig. 162). In 1947 the acreage was 4,600 acres, or 2% of the acreage of all top fruit, and has declined steadily since.

Plums

The acreage under plums in 1970 was 16,033 acres, or 11% of the acreage in commercial orchards. The principal counties are Worcestershire, with 29% of the total acreage, Kent, with 25%, and Cambridgeshire, with 11% (Fig. 163). Provided that drainage is adequate, plums do best on heavy loams, and many of those in the Vale of Evesham were established on the calcareous clays of the Lower Lias. Climatically, the Vale of Evesham is not ideal, since plums tend to split with heavy rainfall; but they are well-suited to the smallholdings which are characteristic of this area. Plums are often grown in market gardens, but are rarely the principal fruit of commercial growers. In 1947 the acreage was 46,600 acres, or 18% of the acreage of all top fruit, and has declined steadily since.

Cherries

There were 7,321 acres under cherries in 1970, or 5% of the acreage in commercial orchards; 80% of this acreage was in Kent, especially in the Faversham–Sittingbourne area, and Worcestershire, with 7%, was the only other county of any importance (Fig. 164). Cherries are intolerant of impeded drainage and require deep well-drained soils,

Fig. 161. Dessert and cooking pears per 1,000 acres of crops and grass.

Fig. 162. Perry pears per 10,000 acres of crops and grass.

Fig. 163. Plums per 1,000 acres of crops and grass.

Fig. 164. Cherries per 1,000 acres of crops and grass.

such as those developed on the brick earths of the North Downs. They are better suited to the drier areas, though they require a good water supply in summer. The concentration of production in Kent was helped by reputation and the existence of a skilled labour force for picking; but problems of disease, difficulties of picking the crop and the long unproductive stage before trees come into bearing have contributed to a decline throughout the 1960s. The acreage in 1947 was 17,600, or 7% of the acreage of top fruit.

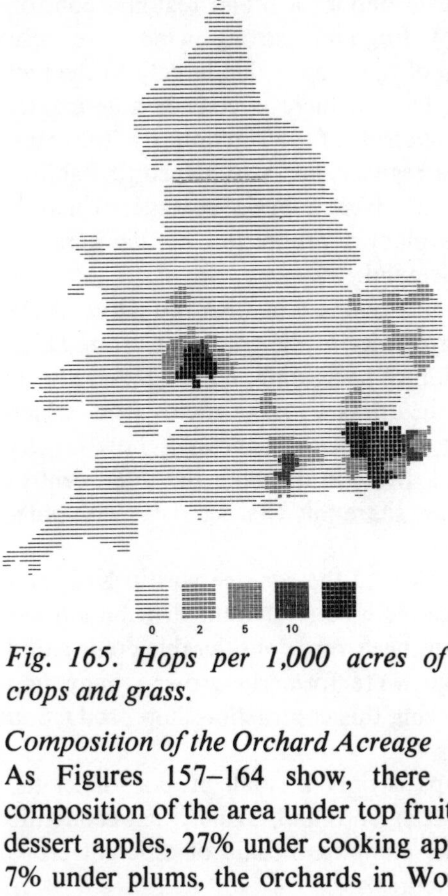

Fig. 165. Hops per 1,000 acres of crops and grass.

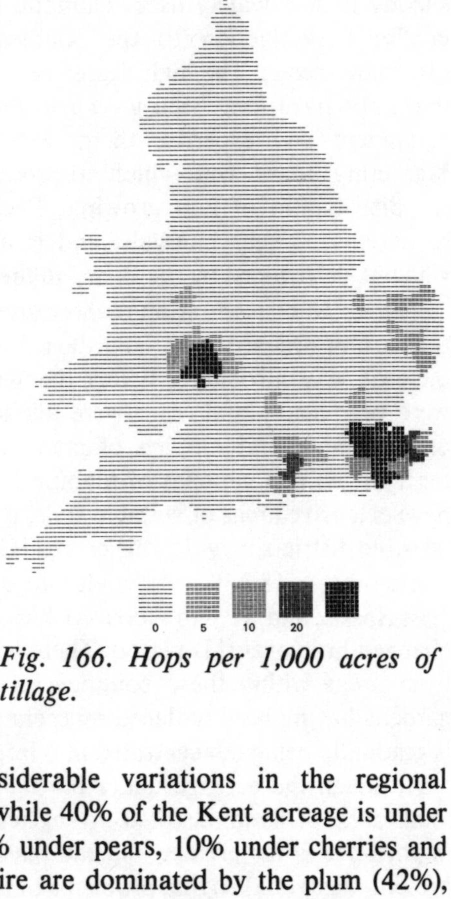

Fig. 166. Hops per 1,000 acres of tillage.

Composition of the Orchard Acreage
As Figures 157–164 show, there are considerable variations in the regional composition of the area under top fruit. Thus, while 40% of the Kent acreage is under dessert apples, 27% under cooking apples, 14% under pears, 10% under cherries and 7% under plums, the orchards in Worcestershire are dominated by the plum (42%), dessert apples accounting for 25% and cooking and cider apples for 9% each. In Devon and Somerset, cider orchards make up 67% of the total, while in East Anglia, dessert apples account for 71%.

Hops

There were 17,493 acres of hops in 1970, or 0·07% of the acreage under crops and grass. The crop is very highly localized, 61% of the acreage being grown in Kent and East Sussex and 35% in Worcestershire and Herefordshire (Figs. 165, 166). Within south-east England hops are grown mainly on soils developed on brick-earth, alluvium and the Upper Greensand.

The concentration is a long-established one, for although hops were grown in Essex, Nottinghamshire and Suffolk in the 18th century, Kent and the west Midlands have long been the leading areas since records were first collected. Hops require deep, well-drained soils, with adequate soil moisture; they are greedy feeders and need considerable supplies of manure. Formerly much of this came from nearby towns, but shoddy is now widely used. Climatic requirements impose a broad regional control, confining production to the southern half of England; strong winds are also disadvantageous. The high degree of localization of the crop is due largely to the fact that hops have long been grown in these areas, so that there is considerable capital equipment and expertise there. Since 1952, control of production by the Hop Marketing Board, with which all producers must register, has also helped to stabilize the distribution of hop growing. Each grower is given a basic acreage, which is reviewed at 5-year intervals, and is allowed to plant a proportion of this acreage, which is determined by the Board in the light of demand.

The initial concentration in these areas is probably due to a number of factors. As in the case of orchard fruit, Kent benefited from proximity to the continent, from which the crop was introduced. Before the mechanization of harvesting, labour requirements were very heavy and there were advantages in nearness to large towns from which casual labour and supplies of manure could be got; but the harvest is now largely mechanized and supplies of manure have ceased. In addition, both the main centres have extensive areas of suitable soils, although they share this characteristic with other possible districts in which the crop might be grown.

There were 18,460 acres under hops in 1938; 64% of the acreage was in Kent and East Sussex and 32% in Herefordshire and Worcestershire, so that the distribution has changed little over this period. There has, however, been very considerable contraction from areas within these counties in which hops were formerly grown, many hop gardens having been replaced by orchards. Following this contraction, hop production is gradually being concentrated into larger units.

Although the acreage under horticultural crops is only a small proportion of the total area of land devoted to agriculture in England and Wales, its economic importance is far greater and the reasons for the complex mosaic of different crops present many challenging problems of interpretation.

CHAPTER VII

Livestock

In this chapter the regional importance of the different classes of livestock will be considered. Livestock are the keystone of farming in England and Wales and more than 80% of all agricultural land is used for their support, as well as large imports of feeding-stuffs. This situation arises partly because climate and relief over much of the country favour a pastoral rather than a cash crop economy, and partly because of the high standard of living enjoyed by the largely urban population of Britain, which places a premium on the production of food rich in protein, such as meat, eggs and milk. Their importance was formerly less marked, but developments in world trade since the mid-19th century, which brought British farmers into competition with those in other, often more favoured lands, led to a concentration of those livestock and livestock products in which the country enjoys the greatest comparative advantage. Sheep and cattle are the principal grazing animals, but pigs and poultry, although they require little land and are being increasingly kept in artificial environments, make an important contribution to agricultural output. Although the kind of livestock varies considerably from region to region, livestock are found in every part of the country, even in those districts almost completely under tillage; on Farm Management Survey farms in the Fenland and on Lincolnshire warpland, for example, 21% of total farm revenue in 1958 was derived from livestock and livestock products. On the other hand, there are many areas where few crops are grown and livestock and livestock products make up nearly the whole of farm revenue.

Despite the importance of livestock, the agricultural census data relating to them are, in many ways, the least satisfactory. Since livestock are often transferred from farm to farm during their lives, so that distributions vary throughout the year, the static picture derived from the June census does not always give an adequate idea of the regional importance of different kinds of livestock; and data from the quarterly censuses are not available by districts. As has been noted, the use of age classes rather than functional types of livestock makes it difficult to do other than hint at important features of livestock farming, e.g., the degree of dependence on other areas for replacements of breeding stock, or differences in the number of lambs born to ewes in various parts of the country. As with crop varieties, it is not possible to take breeds of livestock into account in preparing these maps, although many areas are characterized by distinctive breeds, well adapted to local conditions, and important changes are taking place in their distribution, e.g., the displacement of the Herdwick sheep by the

Swaledale in Cumberland or of the Suffolk sheep by grass and hill breeds, such as the Kerry, in East Anglia.

Before the regional importance of different classes of livestock can be established, they must all be converted into some common unit for comparison, for it is clearly impossible to equate numbers of dairy cattle with those of poultry. Various conversion factors have been devised, mostly based upon the feed requirements of the different classes of stock; those used here, which are based on the Ministry of Agriculture's booklet, *Terms and Definitions used in Farm and Horticultural Management*, are set out in Appendix I.

Figure 167 shows the distribution of livestock units, which represent the sum of all classes of livestock suitably weighted according to their feed requirements. This map indicates the absolute importance of livestock throughout the country, but not their relative importance, which depends on the quality of the land, the suitability of each area for other enterprises and often on tradition (cf. Fig. 219). The greatest number of livestock is to be found in a north–south belt of country, stretching from Cumberland to Dorset, and extending over most of the south-west peninsula. The highest densities are found in south Cheshire and north Shropshire, and in the Fylde, where District 69 has 106 livestock units per 100 acres of crops, grass and rough grazing. Outside this belt there are several areas with high densities, notably Anglesey, south-west Wales and south-east England. The average for England and Wales is 36 livestock units per 100 acres, and densities are generally above this in western England and Wales, except in the uplands where the livestock are mainly sheep and the carrying capacity of the land is low. Densities are also high in the Vale of York, the east Midlands and much of south England. Apart from the uplands, the lowest densities are to be found in eastern England, from the East and West Ridings to Kent; thus in District 188 there are only 8 livestock units per 100 acres of crops and grass and rough grazing.

Cattle are overwhelmingly the most important livestock, accounting for 63% of all livestock units in England and Wales (Figs. 168, 169). They are the leading livestock nearly everywhere in the lowlands, the chief exception being parts of eastern England. Sheep are the second most important class of livestock, accounting for a further 14% and are the leading livestock in upland Wales and the northern Pennines. Pigs account for 12% and predominate over much of East Anglia and in the Fens, and also around the Humber. Poultry (11%) are the leading livestock in only five districts, all in eastern England.

Since ruminants make up 77% of all livestock, this regional distribution of livestock is related primarily to the availability and quality of grazing. Of course, densities cannot be attributed wholly to grazing, for some livestock, particularly pigs and poultry, depend largely on purchased feed, or on arable crops or crop residues, like sugar-beet tops. In a broad view, conditions in eastern counties favour the production of crops and those in western counties favour livestock farming. Where only a small acreage of grass is available or where the quality of the grass is poor, only small numbers of grazing livestock can be carried; where there is abundant grass of good

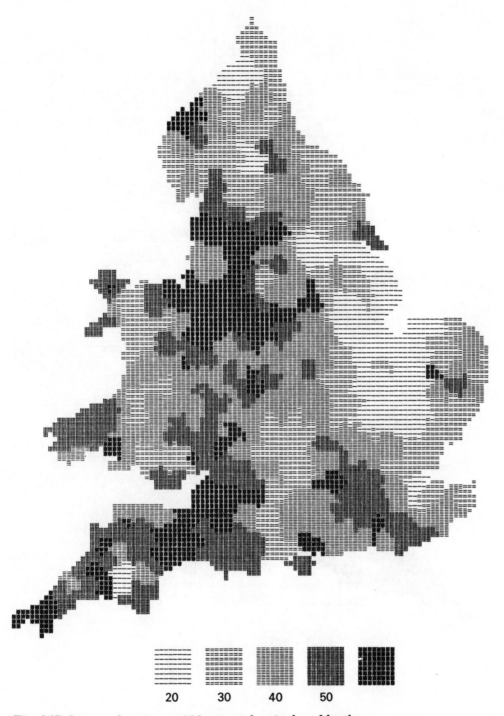

Fig. 167. Livestock units per 100 acres of agricultural land.

167

Fig. 168. Cattle per 100 acres of agricultural land.

quality and the grazing season is long, densities are high. Many of the special reasons for the distribution of livestock will emerge in the subsequent discussion of individual classes of livestock, although the reasons why one class of stock predominates, e.g., cattle on Pevensey Levels and sheep on Romney Marsh, are often far from clear.

Livestock numbered 11,596,827 livestock units in 1970, compared with 9,948,570 in 1938. Cattle then accounted for 55% of all livestock, sheep for 23%, horses for 9% and pigs and poultry for 7% each. The change is primarily due to an increase in the number of cattle, but pigs and poultry were also much more numerous in 1970 than in 1938.

Cattle

The importance of cattle is due primarily to the major role which dairying plays on farms in England and Wales, although beef production, often as a by-product of dairying, has become increasingly important since the Second World War. According to Watson and More, there are few farms which can be run permanently without cattle, and 81% of all holdings of 20 acres and over have cattle.

There were 9,027,703 cattle in 1970, accounting for 83% of all grazing stock (expressed as livestock units), the average density being 33 cattle per 100 acres of crops, grass and rough grazing. Because of the high proportion of livestock units represented by cattle, Figure 168, showing the distribution of cattle, bears many similarities to Figure 167, although the contrasts between different areas are more marked. High values are found principally in Cheshire and Shropshire, although the district with the higher density is again District 69, with 73 cattle per 100 acres of crops, grass and rough grazing. A similar north–south belt of high densities to the east of the Welsh uplands is recognizable. Densities are lowest in the uplands and in Kent, where sheep predominate, and in the Fenland and surrounding country, District 190 having only 3 cattle per 100 acres.

Figure 169 shows the proportion of all livestock units representing cattle in the various districts. The three principal areas in which cattle predominate are: Cheshire, Staffordshire and Derbyshire; east Somerset, Wiltshire and Dorset; and Carmarthen and Pembroke. In most parts of these areas three-quarters or more of all livestock units represent cattle, the highest proportions being 93% in District 350. In most of south and midland England, except the south-west Midlands, Kent and Exmoor, cattle account for more than 60% of all livestock, proportions which are also found in north-west and north-east England and east Lincolnshire. Only in the uplands, south Lancashire, eastern England and Kent are there any large areas where cattle account for less than half the livestock, the lowest proportion being 7% in District 191. All major areas in which cattle are important have a long tradition of cattle rearing, while in most of these districts conditions are not very favourable to arable farming.

Although dairy cattle still consume large quantities of purchased concentrates, they are more dependent on home-produced feed than before 1939, while beef cattle, which consume more roughage, depend largely on the quality of grazing and on home-produced feeding stuffs. The distribution of cattle is thus chiefly related to the availability of feed, particularly grass. Since most cattle are dairy cattle, while even beef cattle do not use the poorer rough grazing, it is the quality and quantity of grass from leys and permanent pasture which is important, and most cattle are to be found in areas with a high proportion of good quality grassland. Feed requirements, however, vary considerably with the class of cattle. The prominence of the south-west is due in part to its lower elevation compared with most uplands elsewhere, to the long growing season for grass and to the prevalence of smaller breeds of dairy cattle in Cornwall. Length of growing season is not the only factor determining the availability of grazing. In low-lying areas of heavy soil the risk of damage by poaching results in a shorter grazing season than the climate alone would allow; on the Weald Clay, for example, it is about 6 months. Summer drought may similarly restrict grazing in eastern counties.

There is some alteration in the pattern of cattle distribution as between summer and winter, reflecting chiefly the seasonal nature of breeding for beef production and the movements of stores and fat cattle discussed below, but it is not often as marked as the seasonal changes in sheep distribution. Winter densities are below those in summer in counties where the breeding of beef animals is relatively important, e.g., Northumberland, or where cattle are fattened on grass in summer, e.g., Leicestershire. They are well above in arable counties, particularly Norfolk, Suffolk and Holland, where animals are fattened in yards in winter.

There were 6,714,261 cattle in 1938, and numbers have risen fairly steadily since; but trends differ, with numbers rising in western and falling in eastern counties, so that while the chief areas remain the same, namely Wales, south, south-west and north England, they had a larger share of the cattle population in 1970 than in 1939, proportions being 51% and 43% respectively. The share of most midland and eastern counties declined; in particular, Cheshire, Derbyshire and Staffordshire, which had 10% of all cattle in 1939, had only 8% of the larger number in 1970.

Dairy Cattle

Milk is the most important single product sold off farms in England and Wales. It is widely produced on farms of all sizes and is often the major enterprise; thus, dairy cattle are found on 52% of all holdings of 20 acres and over. Yet, although the term 'dairy cattle' seems unambiguous, this is not so. A proportion of the cows and heifers are in herds rearing calves for beef, but no clear-cut line can be drawn between the breeding of calves for beef and the production of milk for sale; for example, the Devon, a beef breed, plays an important part in milk production in Devonshire. A considerable proportion of the domestic output of beef, estimated in 1955 at 75%, comes from dairy herds, either as barren or old cows, or as steers and surplus heifers not slaughtered as calves for veal. Nor did the 1970 census distinguish dairy replacements from beef replacements. Figure 171 shows the distribution of dairy cattle in 1970.

170

There are three major areas: north-west England, viz., Cheshire and surrounding counties; Dorset, Somerset and Wiltshire; and south-west Wales. Low densities are found chiefly in the uplands and in eastern England. The highest value is 54 per 100 acres in District 77, compared with an average of 17. Dairy cattle are also relatively most important in these same areas (Fig. 170), although this map must be read with some caution in view of the way dairy cattle are defined.

Dairy Cows

In 1970 there were 2,714,463 cows and heifers in milk and cows in calf (the census

Fig. 169. Cattle livestock units per
100 livestock units.

Fig. 170. Dairy livestock units per
100 livestock units.

categories which are normally taken to represent dairy cows), mainly producing milk or rearing calves for the dairy herd, or 80% of all cows and heifers in milk and cows in calf; this figure may be compared with the 2,688,900 cows estimated to be in milk-selling herds in the Milk Marketing Board's 1970 census. Seventy six per cent of these cows are Friesians, 10% Ayrshires, 9% Channel Island breeds and 2% Dairy Shorthorn. Friesians are the leading breed in every county; only in Kent, Hertford and East Sussex is the proportion less than half, and Wales, the west Midlands and north England have over three-quarters. Ayrshires are relatively more important in north and south England, while Channel Island breeds are most common in southern England, accounting for a third of all cows in Cornwall and Kent.

171

Fig. 171. Dairy cattle per 100 acres of agricultural land.

Fig. 172. Dairy cows per 100 acres of crops and grass.

173

Figure 172 shows the distribution of dairy cows as defined in England and Wales in 1970. There are again three principal areas: Cheshire and the adjacent counties; Somerset and Dorset; and south-west Wales. West Cornwall also appears in the highest category, with 25 or more cows per 100 acres of crops and grass, but the maximum value is 40 in District 77. The lowest densities occur in the uplands, where there are extensive areas of rough grazings incapable of supporting dairy cattle, and in eastern England, where there is little grazing. Although upland Wales and south-east England are somewhat anomalous, there is a broad two-fold contrast between the west, with values above the national average of 12 per 100 acres, and eastern England,

Fig. 173. Dairy cows per 100 cattle.   *Fig. 174. Dairy cows per 100 cows.*

where densities are below average. The distribution of cows in milk-selling herds does not differ greatly from that of cows as defined by the agricultural census.

Figures 173 and 174 show how numbers of dairy cows, expressed as proportions of all cattle and all cows respectively, vary through the country. The main dairying areas again stand out on both maps, with dairy cows representing 40% or more of all cattle and 90% or more of all cows, the highest values being 58% in District 79B and 97% in District 299 respectively. The lowest proportions are to be found in the Welsh uplands, Exmoor and Northumberland, where dairy cows represent less than 20% of cattle and less than 50% of all cows, and in eastern England, especially the Fenland, though in eastern East Anglia and in south-east England proportions are generally above average. High proportions are associated with specialization in dairying,

174

particularly in areas such as the industrial Pennines which depend on others for herd replacements. Low percentages occur in those counties, such as Hereford and Northumberland, where beef cattle are reared in large numbers, and in arable areas where few livestock are kept.

While the distribution of dairy cows does not differ greatly throughout the year, there is some seasonality of production, although it is less marked than in pre-war years. The proportion of summer milk, i.e., that produced in the 6 months between 1 April and 30 September, is higher than in winter, though differences between the highest and lowest monthly totals are much greater. There are also regional differences. Proportions of summer milk ranged from 47% in Norfolk to 60% in Merionethshire; in general, herds in western counties produce considerably more milk in summer than in winter, whereas in several eastern and southern counties slightly more is produced in winter. The chief reasons for this difference are the availability and cost of feed. In the west, where there is little arable land but abundant grass in summer, milk production is cheapest on grass in summer; in the east, where there is less grazing, but more abundant fodder from crops, production is relatively more profitable in winter.

The interpretation of these maps is complicated. The optimum temperature for milk production is about 50°F. (10°C.) so that, from feed and climatic considerations, dairying is chiefly a lowland activity, although it is found at over 1,000 feet in several counties, e.g., Denbighshire. From a physical point of view, the two principal areas, south Cheshire and adjoining districts, and east Somerset, west Wiltshire and Dorset, are both well suited to dairying. They have a high proportion of good grassland, fairly heavy soils and a moderate rainfall, while the south-west has the additional advantage of a longer growing season and mild climate. Largely for these physical reasons, dairying has been long established in the principal areas, although its character has changed considerably.

The present pattern of dairying, however, can be understood only in its historical context. Before the coming of the railway, most milk was made into butter and cheese or fed to livestock, and milk for liquid consumption was produced in town dairies and around the large towns, e.g., in Middlesex. Reductions in the cost of transporting milk and technical improvements in methods of carrying it have gradually extended the area in which liquid milk could be produced for the urban markets, until milk can now travel up to 100 miles by road tanker and even longer distances by rail. The expanding market for milk, arising from the growth of the population, the movement out of arable farming between the 1880s and the 1930s, and the high degree of natural protection which milk enjoys, all encouraged the spread of production for the liquid milk market, both into counties, such as Buckinghamshire, where formerly butter and cheese had been made, and into areas where no dairying of any kind had been practised before. Both milk production and the proportion sold for liquid consumption increased, although there was still a contrast between areas remote from urban markets, where milk was manufactured, or used in livestock rearing, and those which produced milk for sale. The Milk Marketing Board, established in 1933, largely eliminated this distinction by introducing prices for milk which were the same for individual producers irrespective of whether the milk was used for liquid consumption or manufacture, although, as Figure 42 shows, the remoter areas, which tend to concentrate on summer production on grass, supply most of the milk for manufacture into butter, cheese and other milk products. The fixing of regional transport charges for the collection of milk to depots or buyers has also tended to minimize the importance of location. In any case, most of the milk from farms is now sent to large collecting centres in the principal dairying areas and conveyed to urban markets by bulk transport. The chief effect of this pricing policy has thus been to make physical suitability for production of milk a much more important consideration and location of relatively little importance, the reverse of the situation which obtained in the mid-19th century.

Yet it is not wholly true to say that milk production is undertaken in those areas to which it is best suited. Apart from the relative advantages of eastern areas in high cost winter production, farm size is an important consideration. The need for some intensive system of farming, which will produce an adequate income from a small acreage, has made dairying attractive to occupiers of small farms, especially in western areas where intensive arable farming is not possible; the regular income from a monthly milk cheque and the small element of risk in milk production have been added incentives to adopt dairying. Thus, according to the Milk Marketing Board 1970 census, 61% of all herds are on farms of under 100 acres of crops and grass, while nearly three-quarters of all cows are on farms of under 200 acres. In particular, pricing policy and the larger income possible from dairying than from the livestock rearing, which was formerly practised, have encouraged the spread of dairying into small farms on the fringes of the uplands, where physical conditions are often far from ideal and real costs of transport high.

The association of dairying with small farms has become increasingly important in recent years because there has been a growing surplus of milk above the requirements of the liquid milk market. Such milk can be used only to feed to livestock or to make butter, cheese and other milk products which face competition from imports produced more cheaply in other countries. Although Government policy has been to discourage further increases in the output of milk, which are largely due to rising yields and which have the effect of depressing the prices paid to all producers, this is made difficult by the dependence of small farms on milk production, particularly in western and northern counties, and by the relationship of dairying and beef production (p. 170). Nevertheless, partly as a result of price changes and partly through more stringent dairy regulations, there has been a fall in the numbers of both milk producers and small herds. Thus registered milk producers, who numbered 136,520 in 1939 and had increased to 161,890 in 1949, fell to 80,265 in 1970 (though part of this latter fall is due to adjustments in the Milk Marketing Board's register). Similarly, between 1942 and 1970 the number of herds with under 20 cows fell by 75% and the number of cows in them by 68%, while the number of larger herds increased.

Because earlier agricultural statistics do not distinguish between beef and dairy cattle, it is impossible to establish trends over the previous 32 years by examining movements in the number of cows and heifers. Between 1938 and 1970 the number of cows and heifers (both for milk production and beef rearing) rose fairly steadily from 2,611,799 to 3,381,220; but part of this rise, particularly since 1950, reflects the increasing attention being paid to beef production. Milk sales provide an alternative measure, but their use is complicated by the steady rise in milk yields since the Second World War; between 1938 and 1970 sales of milk off farms in England and Wales rose from 1,119 million gallons to 2,257 millions, an increase of 102%, whereas numbers of cows and heifers rose by only 29%. From a consideration of changes in both output and the distribution of cows and heifers, it is clear that there has been a westward shift in the centre of gravity of milk production (although the main dairying areas centred on Cheshire and Somerset remain the leading producers of milk). Between 1957 and 1970 milk sales increased by 58% in the Far Western Region (Devon and Cornwall), and 30% in the Mid Western (Dorset, Somerset and Wiltshire), compared with an average of 21%, while sales in the Eastern region actually declined by 11%.

Regional changes in the number of milk producers also confirm this westerly shift, although allowance must also be made for increases in the average size of dairy herds, which rose from 15 cows in 1942 to 33 in 1970. Between 1957 and 1970, numbers of producers fell in all regions but falls were below average in western regions, especially in Wales, Devon and Cornwall, and above average in eastern regions, notably in the principal arable areas, East Anglia and the Fens, where alternative enterprises are possible.

Herd Replacements

Available agricultural census data do not lend themselves to the analysis of regional differences in the extent to which dairy herds depend on other areas for herd replacements. Only since 1960 have in-calf heifers for dairy and beef herds been separately distinguished, while other female cattle 1 year old and over were not divided into those intended for dairy herd replacements and others until 1971.

Fortunately, this difficulty is less serious than it might at first appear, for dairy cattle do not show marked seasonal differences in their distribution throughout this country, so that this regional distribution of dairy herd replacements corresponds broadly with that of dairy cows. This situation arises partly because breeding is not concentrated in one season and partly because most dairy herds are self-contained units for breeding purposes. Since it is desirable to have a steady supply of milk throughout the year for the liquid milk market, cows do not all calve at the same time, although there are seasonal peaks in autumn and spring; and since most herds breed their own replacements, the seasonal movement of breeding stock, characteristic of sheep rearing, does not take place to any considerable extent, although there is a movement of surplus animals from dairy farms. It is true that some areas do depend on others for breeding stock, but although the rearing of dairy stores on the poorer lands of the upland margins has been shown to be a more efficient use of resources, most farmers, afraid of importing disease, prefer to rear their own replacements.

Figure 175 shows the distribution of first-line replacements for all herds, i.e., in-calf heifers with first calf, the national average being 24 per 1,000 acres. As might be expected from the predominance of dairy cattle and from the dependence of most dairy herds on breeding their own replacements, the distribution resembles that of all cattle, though densities are rather higher than might be expected in north Lancashire and Westmorland. A map of in-calf heifers intended for the dairy herd is very similar, not

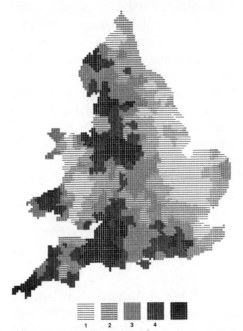

Fig. 175. Dairy heifers per 100 acres of agricultural land.

Fig. 176. Dairy heifers per 100 dairy cows.

surprisingly when 86% of all in-calf heifers are in this category. So, too, is a map of the ratio of dairy heifers to dairy cows (Fig. 176). The national average in 1970 was 21 heifers per 100 cows but proportions were generally lower in the main dairying areas, reaching only 13 in Cheshire; the principal area with a high ratio was north-west England, the average for Cumberland being 27%. Two contributory factors in this pattern are a tendency towards longer herd life in some areas such as Cheshire and dependence on other areas for replacements; thus the Lake District and the dales of the northern Pennines supply herd replacements for the dairy farms in the industrial areas further south where cows are often sold after two years in the dairy herd.

In many of the areas with high values in eastern England the numbers are small and, for the same reason, a map of the ratio of in-calf heifers intended mainly for the rearing of calves for beef to cows in the beef herd is not very meaningful. Ratios of heifers to beef cows are, however, generally lower, averaging 15; for the herd life of such cows is usually longer than that of dairy cows. A further problem of interpretation arises from the considerable regional differences in the percentage of heifers calving in each quarter. These are partly due to differences in the seasonality of calving of beef and dairy cows; thus Wales, the south-west and north-west, where breeding of beef cattle is often important, tend to have low proportions of heifers calving in the September–November quarter, while areas such as Norfolk, where there is an emphasis on winter milk production, tend to have high proportions.

Beef Cattle

The breeding, rearing and fattening of beef cattle are less important enterprises than the keeping of dairy cattle and are rarely the major enterprise on farms, except in a few areas. Indeed, as noted, much of the beef produced is the by-product of dairy farming. Consequently the interpretation of what are beef cattle is even more difficult than in the case of dairy cattle. It can be safely assumed that all male cattle other than bulls and bull calves, whose number is small and declining as a result of the increasing importance of artificial insemination, are kept for beef, although many of these animals originate from dairy herds. Cows producing calves mainly for beef can also be allocated to beef production, as can that proportion of the younger female cattle which are not required for herd replacements. With the increasing use of beef bulls on dairy cows, many of these heifer calves will be crosses. Figure 177 shows the distribution of such cattle, which averaged 17 per 100 acres, and reveals three main areas, viz., south-west England; the Welsh borderland, notably Herefordshire; and north-east England. The highest density, 44 per 100 acres, was in District 302A, in Anglesey, and other areas with above-average values include the Solway lowlands; north-west and south-west Wales; the east Midlands; Buckinghamshire and Sussex. The lowest values are in East Anglia where numbers of all grazing livestock are small, although it should be remembered that this is a summer distribution and that densities are highest in winter when cattle are sent here for fattening in yards on arable crops.

The production of beef cattle falls broadly into three stages: breeding, rearing and fattening, each of which has its own requirements and tends to be more common in certain parts of the country, although all these stages are widespread and may well be found on the same farm. As with dairy cattle, there are several breeds of beef cattle which also tend to be concentrated in certain parts of the country, e.g., Aberdeen Angus in Northumberland. But many cattle fattened for beef are of dairy or dual purpose breeds, or are crosses of beef bulls and dairy cows; similarly, not all the cows and heifers mainly producing or rearing beef calves are of beef breeds.

Breeding
In 1970 there were some 666,757 cows and heifers in milk and cows in calf mainly

179

Fig. 177. Beef cattle per 100 acres of agricultural land.

producing or rearing beef calves, or 20% of the total number of the breeding herd. Such cows are to be found chiefly in three areas, in each of which there are 5 or more cows per 100 acres of agricultural land, compared with a national average of 2: Northumberland and West Durham; upland Wales and the Welsh borderland; and the uplands of south-west England (Fig. 178). The highest value is 11 in District 321. Other important areas are the Lake District, the Pennines, east Lincolnshire and West Sussex, but some beef cows are found in all districts, with lowest densities in the main arable counties of eastern England.

Fig. 178. Beef cows per 1,000 acres of agricultural land.

Fig. 179. Beef cows per 100 cows.

There are also considerable regional variations in the proportion of cows which are kept mainly for rearing beef animals (Fig. 179). Proportions are highest in the three areas already noted, in upland Wales, upland Northumberland and Durham and Exmoor, with proportions of 50% and over; but high proportions also occur in eastern England, especially around the Fenland, where densities are low, and Lindsey, where beef cows are much more numerous. The lowest percentages are to be found in the main dairying areas, with values everywhere of less than 10%. This map is, of course, the mirror image of Figure 174.

Reasons for this distribution are in part physical and in part economic and social. Animals being reared for beef, particularly those of hardy breeds such as the Welsh Black and the Aberdeen Angus, are less demanding in their requirements than dairy cattle and are consequently able to make use of the poorer permanent grassland and of the lower rough grazings. Since the rearing of beef cattle has not in the past been a particularly profitable enterprise, it tended to be located on low-rented land. Because the young stock require considerable attention, breeding and rearing have often taken place on the small family farms on the upland margins, although, as has been noted, these were often the farms which changed from livestock rearing to dairying in the 1930s and 1940s. The breeding of beef calves is, of course, not confined to the upland margins; it occurs in lowland Hereford, in Devon and Cornwall and on large farms in Northumberland. Why there should be this concentration only on parts of the upland margins is not clear; in part it is due to tradition, for Devon and Herefordshire are both the home of a breed of cattle which bears the county name and predominates in that county.

Although there is this regional specialization in breeding, its importance should not be exaggerated, for steer calves from dairy herds make up a large proportion of steer calves reared for beef. Over a third of all artificial inseminations, which are now the principal way of mating dairy cattle, are from bulls of beef breeds, and such breeding takes place in the main dairying areas, although the calves are soon moved elsewhere.

Data on numbers of beef cows, available since 1954, show an increase of 73% by 1970. It has been Government policy since the 1940s to stimulate the production of beef, through continuation of the Hill Cow subsidy, first paid in 1943; a Beef Cow subsidy, introduced in 1966 to encourage beef herds on poor ground in the lowland; a calf subsidy on 8-month-old calves (6 months in hill country) suitable for beef production; and partly by improvements in guaranteed prices for fat cattle. The growing interest in beef cattle is also shown in the rise in the percentage of all artificial inseminations of dairy cows which are from beef bulls, which increased from 19% in 1953 to 38% in 1970.

Rearing

Figure 180 shows the distribution of steer calves under 1 year old in June 1970, when they numbered 1,215,780. These animals consist of two groups, those from beef herds and those from dairy herds. Steers born on dairy farms may be slaughtered for veal, they may be reared and fattened for beef, or they may be sold as calves or as store cattle to be fattened elsewhere. Steers from beef herds, on the other hand, are generally reared, at least for some months, on the farms where they were born. Steer calves are most numerous in four areas: the margins of the Welsh uplands, especially in the Welsh borderland; Cumberland; the Vale of York and the East Riding; and south-west England. The highest value recorded is 105 steers per 1,000 acres of agricultural land in District 33, compared with an average of 43. Other important areas are Northumberland, the east Midlands, Buckinghamshire and Sussex. Low values are associated with areas of dairy farming, where many are born but few retained, or intensive horticulture, where there are few grazing livestock, and with high moorlands, where conditions are too harsh for cattle.

182

Fig. 180. Steers under 1 year old per
1,000 acres of agricultural land.

Fig. 181. Steers under 1 year old per
100 cows.

Fig. 182. Steers 1 year old and under
2 per 1,000 acres of agricultural land.

Fig. 183. Steers 1 year old and under
2 per 100 cattle.

183

Figure 181 gives some idea of the extent to which steer calves are retained on the holdings on which (or more strictly the districts in which) they were born. If all steers were reared on the farms of birth the distribution of steers would resemble that of cows, if the effects of any difference in seasonality in calving are disregarded. Yet there are considerable regional differences in the ratio of steer calves to cows. Proportions are lowest in the three main dairying areas, reaching a minimum of 7% in District 347A; they are highest in the East Riding, the Vale of York, the Fenland and parts of East Anglia, with 6 steer calves for every cow in District 189, and are generally higher in eastern counties. The low percentages in the main dairying areas reflect both the small proportion of cows rearing mainly beef calves and the fact that most steer calves are soon sold. Yet, even among dairy herds there is considerable regional variation in practice. According to the Milk Marketing Board's 1960 census, the most common practice in the main dairying areas is to dispose of steer calves shortly after birth, whereas in many herds in northern England calves are reared to yearling stage, and in the east Midlands beef calves are most commonly reared to maturity. The high proportions of steer calves in eastern districts show that they move here from other areas, although, in the case of the Fenland, the actual numbers are small. Such animals are generally moving to better quality land, or to areas where feed is available or where they do not compete with dairy cattle.

Numbers of steer calves have been rising steadily since 1947, when there were 278,835, for rearing has been more profitable. In the period of which data are available they have been increasing at a faster rate than those of cows, indicating that a large proportion of these steer calves must come from dairy herds. This trend is also confirmed by the fact that a larger proportion now comes from dairying counties, representing animals which would formerly have been slaughtered for veal, although the largest proportionate increase has occurred in the southern group of counties between Dorset and West Sussex. The share of all the principal areas, except the East and North Ridings, has fallen; thus in 1942, 14% of all steer calves were in Devon and Cornwall and 20% in Wales and Hereford, compared with 10% and 16% respectively in 1970.

Once they have passed the age of 8 months or so, steers require little attention and are kept as store cattle in all parts of the country. There were 798,532 steers between 1 and 2 years old in 1970. Figure 182 shows their distribution, which resembles that of steer calves; the chief areas are the Vale of York, Cumberland and Northumberland, Hereford, north-west and south-west Wales, and Cornwall, the highest value being 91 per 1,000 acres of agricultural land in District 302, compared with an average of 29. How many of these are stores and how many are being fattened for slaughter before they are 2 years old cannot be determined from the census, but it seems likely that many such steers in areas which are important for the fattening of older cattle are being fattened for slaughter (Fig. 184). There were 470,215 steers between 1 and 2 years old in 1940, but the number has been rising steadily since 1948, when there were 276,086. They, too, have tended to become dispersed more widely throughout England

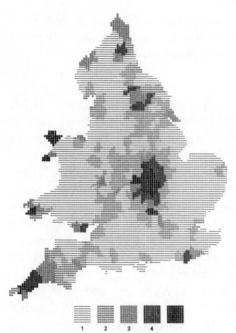

Fig. 184. Steers 2 years old and over per 1,000 acres of agricultural land.

Fig. 185. Steers 2 years old and over per 100 cattle.

Fig. 186. Cattle being reared intensively per 1,000 acres of agricultural land.

Fig. 187. Cattle being reared intensively per 100 other cattle.

and Wales. Figure 183 shows steers between 1 and 2 years old as a proportion of all cattle. The highest values, exceeding 15%, are in eastern England, though numbers are often small in such areas; values are also above average (10%) in the main areas.

Fattening

Cattle for fattening are not distinguished as such in the agricultural statistics. As has been noted, cattle for beef come from a number of sources. Culled dairy cows make an important contribution, estimated at a fifth; unfortunately, they cannot be distinguished from dry cows and the dairy herd replacements which are also included in the category 'Other female cattle 2 years old and over'. Steers and heifers, largely from dairy herds, supply most of the animals for fattening, but these can be slaughtered at varying ages; in recent years the average age of slaughter has been falling and this trend implies that some animals for fattening have already been included in the map showing steers of 1 year old and under 2 (Fig. 182). There is, however, no doubt that the distribution of male animals 2 years old and over, excluding bulls, gives a good indication of areas of summer fattening.

In 1970 there were 333,593 other male cattle 2 years old and over. Their distribution (Fig. 184) has many features in common with that of steers 1 to 2 and of Irish cattle. The principal area is the east Midlands (Leicestershire and Northamptonshire), but Anglesey and Cornwall are also important; the highest value is 114 per 1,000 acres of agricultural land, compared with an average of 12. Densities are lowest in the uplands, the main dairying areas and the principal arable areas. The map of steers 2 years old and over as a proportion of all cattle (Fig. 185) shows substantially the same pattern as Figure 184, although eastern areas are more prominent. The average value is 4% and the maximum 23% in District 147.

Summer fattening of cattle is undertaken primarily on grass. Cattle cannot be fattened on the poor grazings of the uplands on which some of them were reared, nor can they generally compete with dairy cattle in the main dairying areas. In Cheshire, for example, the pastures are said to favour milk secretion rather than liveweight increase. Some cattle are fattened in most lowland areas, even if the land is not very suitable; it has been shown that cattle fattened in Wales achieved only half the liveweight increase (considering only gains from grass) of animals fattened on the high quality grassland of the east Midlands. The major cattle fattening areas are mainly of two kinds; areas where there are high quality pastures, often on fairly heavy land, and coastal areas where there is abundant growth of grass. Yet it is difficult to find adequate reasons to explain their distribution, although tradition and reputation seem to play an important part. The pastures of the east Midlands are reputed to yield a higher return from beef fattening than from dairying and the fattening pastures are of high quality and have certainly been long established. On the marshes there is a good growth of grass and few alternative enterprises are possible.

Cattle are also fattened in winter. In the traditional arable economy of eastern England, yarded cattle, fed on roots and straw, provided abundant farmyard manure for the arable land; indeed, potato growers in the Fenland and Brussels sprouts growers in Bedfordshire kept such cattle largely for the muck they provided. The practice of buying in stores for winter fattening is less common in the arable areas than formerly, but it is still widely practised. The explanation of this situation lies partly in the widespread belief among farmers that this is the proper way to farm and partly in the pleasure that many farmers get from keeping beef cattle, although winter fattening often brings little financial gain and may even result in a loss. In most arable areas numbers of steers are considerably higher in winter, but even a comparison of the June and December censuses is misleading, for it is common practice to fatten cattle in yards in two batches, one being cleared by January, the other in the spring. By contrast, in most of the traditional areas of summer fattening, there are fewer older steers in winter.

Methods of cattle fattening, whether on grass or on arable farms, have tended to become more intensive as rents have risen and profit margins have been squeezed; the most extreme form is the production of baby beef, in which calves, often Friesians, are fed on a diet of concentrates. Such intensive systems are commonest in the main areas of arable farming, especially Yorkshire, though some are also found in the main dairy areas (Fig. 186). There were 194,612 steers and 30,320 heifers being reared intensively in 1970 (a reminder that the preceding maps omit heifers and cows being fattened). Intensively reared cattle are relatively most important in Yorkshire and East Anglia, where they account for 10% or more of all other cattle, that is, excluding the breeding herd (Fig. 187).

England and Wales are not a self-contained unit for livestock farming and there is considerable movement to and from Scotland and Ireland, though store cattle from Ireland are the only category which can be mapped. Irish cattle numbered 101,255 in June 1970, or 2% of all other cattle. They average 4 per 1,000 acres of agricultural land and are most numerous in Anglesey, the east Midlands, the Vale of York and Northumberland, with 20 or more (Fig. 188), although only in Wales and the southwest are there large areas without any. Figure 189, showing Irish cattle as a percentage of all other cattle, is very similar. It should be noted that all the cattle recorded in Figures 188 and 189 have already been included with the appropriate age group in Figures 182–185.

In 1940, when they were first separately recorded, there were 446,845 male cattle 2 years old and over, or 6% of all cattle. Numbers fell during the Second World War, but rose to a maximum of 560,885 in 1956. They did not increase at the same rate as younger steers and steer calves during the 1950s and have tended to decline in the 1960s, confirming the trend to slaughtering at younger ages. The principal areas in 1940 were the same as in 1970, but, while Anglesey, Northumberland and Cornwall had a slightly larger share, the contribution of the east Midlands was less important; thus, the share of Leicester, Northampton and Rutland declined from 15% in 1940 to 13% in 1970. The share of East Anglia and of counties along the Welsh border has also fallen.

Sheep

Sheep are far less important than cattle on farms in England and Wales. In 1970 they were absent from 72% of all holdings, compared with 34% in the case of cattle; they are even less common on small holdings, and there were no sheep on 85% of all holdings with under 50 acres of crops and grass. Yet, paradoxically, in those areas where sheep are numerous they are often the only or the major enterprise.

The pattern of sheep farming is extremely complex and available statistics do not lend themselves to the close analysis of regional differences. The main aim of sheep

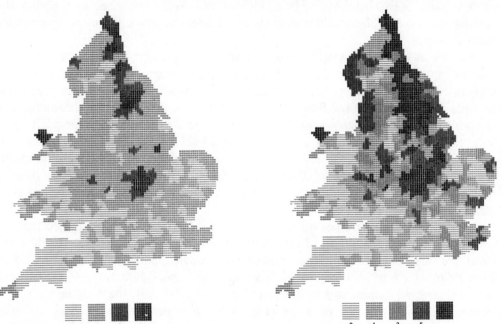

Fig. 188. Irish store cattle per 1,000 acres of agricultural land.

Fig. 189. Irish store cattle per 100 other cattle.

farming in England and Wales is the production of fat lambs. For biological reasons this is a markedly seasonal activity; but the periods at which lambing, disposal and slaughter take place vary considerably throughout the country. Furthermore, there are complex seasonal movements of stock, mainly from uplands to lowlands, but also within the lowlands and, on a smaller scale, from lowland to upland, and here, too, there are considerable regional differences in the dates at which these events occur. To interpret such changes from data referring to a fixed date in summer is, therefore, fraught with difficulty.

As with cattle, there are many different breeds of sheep, numbering over 40. Many of these are adapted to particular environments, particularly the hill breeds, and some features of the distribution of the sheep population need information about the

188

Fig. 190. Sheep per 100 acres of agricultural land.

prevalence of different breeds for their proper interpretation. No detailed source, however, is available.

In June 1970 there were 17,619,773 sheep and lambs in England and Wales, or 13% of all livestock (in livestock units), the average density being 57 per 100 acres of crops, grass and rough grazing. As Figure 190 shows, sheep are largely confined to the uplands, the highest densities occurring in upland Wales. Exmoor, the northern Pennines, Northumberland and much of Wales and the Welsh borderland have one sheep or more to the acre; the only upland in which sheep are relatively few is the southern Pennines, which lie between the industrial towns of Yorkshire and Lancashire. All northern England except the vicinity of the Tyneside conurbation, all Wales except coastal Pembroke and Carmarthen, and all Devon and Cornwall except the south-west tip, have densities of 50 or more per 100 acres. Such densities are matched in only two of the remaining lowland areas: Kent, where Romney Marsh has 193 per 100 acres, and the south Midlands (Gloucestershire and Northamptonshire). Sheep are least numerous in the main arable areas.

It must be stressed that this is a summer distribution. Numbers of sheep in winter are appreciably lower because of the slaughter of most of the wether lambs born in the preceding spring; thus in December 1969 there were 12,748,000 sheep and lambs, or 73% of the June total, although there are also regional differences. These are due to a variety of factors, such as the size of the lamb crop, the age at which lambs are sold, the incidence of mortality among sheep and the number of ewe lambs retained for breeding purposes.

Figure 191 shows the proportion of all livestock (in livestock units) which represents sheep. It strongly resembles the map of the distribution of sheep and confirms that sheep account for the highest proportion of livestock where they are most numerous. In all the major uplands, a third or more of the livestock are sheep, compared with an average of 13%, the highest proportion being 68% in District 316A. Considerable higher proportions would have been recorded if districts in the uplands did not also include adjacent lowlands. The lowest proportions are found in East Anglia and the main dairy farming areas. By way of contrast, a map of the proportion of livestock which represents beef cattle is shown alongside (Fig. 192). This shows that over much of the uplands beef cattle and sheep are not in competition; it also illustrates the greater importance of beef cattle in the lowlands, especially in the Midlands.

Interpretation of this distribution of sheep involves physical, economic and historical considerations. Although the sheep is best suited to well-drained pastures on chalk and limestone, it is better able to survive under the rigorous conditions on the uplands than any other class of livestock; indeed, the higher uplands could be farmed in no other way. Despite the severe winters, the heavy rainfall and the damp, windy weather, sheep can breed and rear lambs. They can utilize the poor rough grazings with little, if any, supplementary feeding, and even in the lowlands derive most of their feed from grass. Moreover, they require little in the way of fixed equipment. The use of the uplands to provide breeding stock and store lambs for the lowlands may be regarded as an attempt to maximize the product of national resources by using the poorer land for the least profitable branch of sheep farming. Ewes of hardy breeds from the uplands, crossed with rams of lowland breeds, also help to provide prolific crops of lambs of kinds required by the market. Differences in density between the various uplands are due largely to differences in the quality of grazing and in the breed of sheep kept. Densities are highest in Wales because the principal breed, the Welsh Mountain, is a small sheep which can be relatively densely stocked and because the quality of the grazing is better than elsewhere and more use is made of the fridd or in-bye land. Lower densities in northern England reflect both the more northerly latitude

Fig. 191. Sheep livestock units per 100 livestock units.

Fig. 192. Beef livestock units per 100 livestock units.

and the fact that most sheep are of larger framed breeds, e.g., the Swaledale. In parts of both the Pennines and Wales densities are rather higher than the land justifies as a result of overstocking the common rough grazings.

Although it is possible to provide explanations of the present distribution of sheep, it must be remembered that this concentration of sheep in the uplands is of comparatively recent origin. A map of 1875 would have shown densities on the clay lowlands of the Midlands nearly as high as those in the uplands, while throughout the arable counties densities would have exceeded 50 sheep per 100 acres. There are numerous reasons for the disappearance of sheep from many parts of the lowlands: high labour costs of managing folded sheep on arable land in a period of acute depression in arable farming; the expansion of dairy farming, with dairy cattle competing for available grass; the growth of towns and their associated population of dogs which discourage sheep-keeping near towns; consumer preference for smaller joints, which favours hill rather than lowland breeds; competition of refrigerated meat from other countries, especially New Zealand; difficulties of obtaining shepherds; and the policy during the Second World War of favouring cattle at the expense of all other classes of stock. Paradoxically, sheep in the lowlands are chiefly found on grassland, where they are liable to foot rot and liver fluke, but where they are more easily managed than on arable land and can, up to a point, complement the grazing of cattle, giving a higher output than from either sheep or cattle alone. It was formerly held that folded sheep were essential to the maintenance of fertility on light arable soils, but this has been shown not to be the case. Sheep are still found on arable farms, where they perform a useful role as scavengers, eating up crop residues, but their numbers are now comparatively small. The survival of sheep in the Midlands is partly a consequence of the quality of the grassland and partly a matter of tradition. In Romney Marsh, too, sheep rearing has survived partly out of ingrained habit. This area, unique in the lowlands, is able to sustain high densities of sheep by virtue of its specialization and the high quality of its pastures. In many ways it fulfils a similar role to that of the uplands, acting as a reservoir of breeding stock and a source of store sheep for fattening elsewhere.

Numbers of sheep have fluctuated greatly over the 32 years since 1938, when there were 17,912,508 sheep. Numbers fell during the Second World War and reached a minimum of 10,161,609 in 1947, following the severe winter of 1946–7, which resulted in heavy mortality among sheep on upland farms. Numbers rose steadily to a peak of 20,525,912 in 1966 but have since declined. In 1939, also, sheep were found mainly in the uplands, 26% being in Wales alone; during the war, the degree of concentration became more marked, the percentage in Wales reaching 32% in 1946. It fell again in the immediate post-war period, but is now higher than ever at 34% in 1970. The proportion of the sheep population in the eastern counties fell during the war and has not risen since; the percentage in other lowland counties also fell, but has increased since, although it remains lower than in 1939.

Details of the regional character of sheep farming are difficult to map and Figures 193–196 have been included to draw attention to important aspects which cannot be adequately analysed with available data. Figure 193 shows the distribution of the breeding flock in June: as would be expected, densities are highest in the principal sheep rearing areas, although the disposal of some flying flocks before the June census may minimize the importance of breeding in the lowlands, while drafting of ewes and winter deaths reduce that of the uplands; in some parts of the uplands only three-quarters of the ewes survive to be drafted out after 4 years. The number of shearling ewes as a percentage of breeding ewes is an indication of flock replacements (Fig. 194). The highest proportions occur in the uplands, where flocks provide their own replacements and ewes are commonly kept for three or four seasons before being drafted to farms on lower ground to provide a crop of cross-bred lambs. Lowland

Fig. 193. Ewes per 100 acres of agricultural land.

Fig. 194. Shearling ewes per 100 ewes.

Fig. 195. Wethers per 1,000 acres of agricultural land.

Fig. 196. Lambs per 100 sheep.

areas, on the other hand, commonly rely on the uplands for their replacements of breeding stock and consequently show generally lower values (though there are many anomalies, which can probably be explained by the very small numbers of sheep in many lowland districts).

Other sheep 1 year old and over are presumably mainly wethers, although some draft or cast ewes may erroneously be returned in this category. Densities are highest in Kent, in upland Wales, on Exmoor and in the Lake District, with 30 or more per 1,000 acres of agricultural land; the highest density is 121 in District 328 and the average 12 (Fig. 195). Explanations of these differences are to be sought largely in breed of sheep and availability of feed. The very high densities in Kent are due to the fact that the progeny of the Romney Marsh breed are slow maturing and are generally sold fat when about either 1 or 1½ years old. In Wales and the Lake District, small wether lambs, which will not get too large for the market, are probably retained to consume surplus grass in summer. The true wether, the policeman of the upland sheep walks, has largely disappeared as a result of changes in consumer demand, from mutton to lamb and from large joints to small.

Owing to the climatic differences in the main sheep rearing areas there are wide variations in the numbers of lambs born per 100 ewes. Unfortunately it is not possible to map these variations in the lambing percentage because both ewes and lambs are being drafted out of flocks at different times. Ewes of the Dorset Horn breed lamb in autumn and most lowland ewes lamb early in the year, so that their offspring are often disposed of before the June census; in the uplands, on the other hand, lambing takes place in March and April, or even May. In Devon, for example, on a sample of better land farms 42% of lambs were born in January or February and 49% in February or March, while on moorland farms 87% were born in March or April. Ewes on low ground in Breconshire produce between 110 and 150 lambs per 100 ewes, while on the margins of the hills the lambing percentage is between 80 and 85, and on the hill country between 60 and 65, compared with 160–70 in true lowland areas. Mortality rates are also probably higher on the uplands. Figure 196 gives some indication of these differences by showing that in the uplands, where few, if any, of the lambs have been drafted out by early June, sheep 1 year and over (mainly breeding ewes) are more numerous than lambs; in the Lowlands, there are more lambs than adult sheep except where late maturing breeds are kept and there are many wethers, or where considerable numbers of lambs have been sold before June.

Pigs

The rearing and fattening of pigs for pork and bacon is an intensive form of production which is little related to the nature of the land itself. The pig eats little grass and is largely dependent on concentrates or on products such as surplus or sub-standard potatoes. It is also a prolific animal and numbers are capable of varying greatly in a short period. In addition to long-term trends, pig numbers undergo cyclical

Fig. 197. Pigs per 100 acres of crops and grass.

fluctuations, referred to as the pig cycle; as numbers increase, prices are reduced, numbers fall and prices begin to rise. For this reason, it can be misleading to examine the pig population in only one year; in June 1970 pig numbers, particularly those of breeding sows, were rising. There is, on the other hand, no reason to suppose that the cycle greatly affects the relative importance of different areas.

In 1970 there were 6,408,391 pigs in England and Wales, 14% of all livestock, the average density being 25 per 100 acres of crops and grass or 23 per 100 acres of crops, grass and rough grazing. Pigs are most numerous in four areas: around London, especially in the horticultural districts; around the south Lancashire and west

Fig. 198. Pigs per 100 acres of agricultural land.

Fig. 199. Pig livestock units per 100 livestock units.

Yorkshire conurbations; the eastern half of East Anglia; and Holderness (Fig. 197). In all these districts there are 50 or more pigs per 100 acres of crops and grass, the highest density being 143 in District 175. Pigs are fairly numerous throughout the lowlands, the chief exception being the livestock areas of the east Midlands. Pigs are unimportant in and around the uplands, whether densities are measured by references to the crops and grass acreage or total agricultural land (Fig. 198). In fact, there is surprisingly little difference between the two maps, emphasizing that pigs are essentially a feature of the farming of lowland England.

Expressed as a percentage of all livestock, pigs have a somewhat different regional importance (Fig. 199). The highest proportions are to be found mainly in areas of intensive vegetable or cash root production, where livestock densities are low and there is little grazing but often considerable quantities of crop residues; in District 189 pigs, as livestock units, account for 63% of all livestock. In East Anglia, the Fens, Holderness and the Humber warplands pigs account for 30% or more, while in the Home Counties, the West Riding and south Lancashire, proportions are 20% and over. Proportions are somewhat lower in the principal dairying districts, while pigs are both absolutely and relatively unimportant in the uplands.

As Figures 200 and 201 suggest, there is relatively little regional specialization in pig breeding and fattening, although there is more specialization between holdings; for example, a sample survey in 1960 by the Pig Industry Development Authority showed

Fig. 200. Breeding sows per 100 acres of agricultural land.

Fig. 201. Other pigs per 100 acres of agricultural land.

that 36% of herds in Great Britain were concerned with breeding only and 16% with feeding only. Both breeding sows and young pigs are most numerous in the same areas, although other pigs are relatively more important around the towns than elsewhere; in Middlesex and around Manchester there are more than 10 other pigs to every sow, and in much of south Lancashire and north Cheshire over 8, compared with the national average of under 7.

The distribution of pigs is related partly to physical conditions, but it is primarily to be explained by the structure of farms and holdings, by the availability of food and by past farming systems. The pig is the most sensitive of all farm animals to cold and damp and consequently is largely excluded from upland areas. Yet the importance of external climate must not be exaggerated, for pigs are often kept in artificial environments; the 1960 sample survey of herds showed that rearing sows were housed all summer in 69% of herds and all winter in 78%. Availability of feed is also important, particularly around large towns and in areas like East Anglia (barley) or the Fenland (potatoes). Yet the principal factor is probably size of holding; for pigs, as an intensive form of production, are often associated with small holdings. Thus, 65% of all holdings which had pigs in 1970 were under 100 acres of crops and grass in extent and these holdings contained half the total number of breeding sows. Indeed, the map, by referring to holdings of 1 acre and above, minimizes the importance of small holdings; in Lancashire, for example, it was estimated in 1960 that 1,400 holdings under 1 acre had 24,000 pigs, or about a tenth of those in the country.

The origins of pig-keeping in several of the more important areas are to be sought in systems of farming which are no longer practised. Thus, in Cornwall the supply of skimmed milk from butter and cream making on farms was an incentive to keep pigs, although the small size of most holdings in the county must also have been important. Similarly, the making of farmhouse cheese in Cheshire, a practice which has virtually disappeared, encouraged the keeping of large numbers of pigs.

The number of pigs has changed considerably over the 32 years since 1938, when the pig population was 3,564,267. During the Second World War, numbers were drastically curtailed to save imported feeding-stuffs, and fell still further after the severe winter of 1947 to a minimum of 1,145,898. Since then, the trend has been upwards, even though prices fell in the 1950s, and numbers in 1970 were nearly 85% higher than those in 1938. The distribution of pigs has not changed greatly over this period, although numbers have increased everywhere except in central Wales. Most counties now have a rather larger share of the pig population, while that of Wales and south-west England fell from 26% to 20%.

Poultry

The keeping of poultry, both for the production of eggs and for meat, is, like pig keeping, an intensive form of agricultural production in which the nature of the land plays little part. Numbers of poultry are similarly capable of great fluctuations, while the character of the industry has changed markedly over the past two decades.

There were 116,681,550 poultry of all ages in 1970, or 11% of all livestock (in livestock units), an average of 482 per 100 acres of crops and grass or 430 per 100 acres of crops, grass and rough grazing. Although poultry are found in all areas, their distribution is extremely uneven (Figs. 202, 203). As a broad generalization, densities are highest in south Lancashire and the West Riding; eastern East Anglia; and the Home Counties, but there are many isolated districts with values of 1,000 birds or more per 100 acres, whether of agricultural land or of crops and grass. The highest value, 7,524 per 100 acres of crops and grass, compared with an average of 482, occurs in District 169. Other areas where poultry are important include Teeside, the Vale of York, Lindsey, Nottinghamshire, the south-west Midlands and Hampshire.

Fig. 202. *Poultry per 100 acres of agricultural land.*

The east Midlands are the principal lowland area with relatively few poultry, while densities in the uplands are nearly everywhere under one bird per acre.

When the proportion of livestock units represented by poultry is mapped, the resulting pattern is broadly similar to those shown by the distribution maps (Fig. 204), with poultry accounting for at least a fifth of livestock in the main areas; the eastern half of East Anglia is the chief exception. Poultry accounts for less than 5% over most of the uplands.

The term 'poultry' covers a variety of birds, although 95% of the total in 1970 were fowls; 4,600,785 turkeys (3·9%), 1,185,795 ducks (1·1%) and 152,505 geese (0·1%)

Fig. 203. Poultry per 100 acres of crops and grass.

Fig. 204. Poultry livestock units per 100 livestock units.

make up the balance. Fowls kept for egg production account for 53% of all poultry, but there are noteworthy differences in their distribution (Fig. 205), with Lancashire and the West Riding as the principal area, and East Anglia of secondary importance; the highest value is, however, in Carmarthenshire (District 347A, 1,539 per 100 acres). A considerable part of the laying flock is to be found around the main urban centres. The distribution of fowl kept for breeding is much more dispersed throughout the lowlands. So, too, is the distribution of broilers, though Lancashire and the West Riding are much less important than they are in egg production (Fig. 206). Production of other table birds (nearly forty times less numerous) is similarly concentrated in a relatively small number of districts scattered throughout the lowlands. Turkeys are characteristic of Norfolk and Yorkshire, which have almost half the flock (Fig. 207).

Fig. 205. Fowls in laying flocks per 100 acres of agricultural land.

Fig. 206. Broilers per 100 acres of agricultural land.

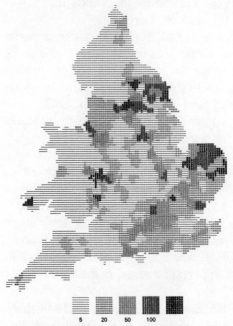

Fig. 207. Turkeys per 1,000 acres of agricultural land.

Fig. 208. Ducks per 1,000 acres of agricultural land.

201

Ducks are to be found mainly in Norfolk, and geese are more characteristic of the upland margins (Figs. 208, 209).

The reasons for regional specialization in poultry-keeping and in the different classes of poultry are many and varied. Like pigs, poultry are often kept in artificial environments and the proportion so kept is increasing; thus, while 95% of laying fowl in 1948 were kept under the range system, by 1970 the proportion had fallen to 5%, as increasing numbers of birds were kept under intensive systems of management. Poultry-keeping has often been associated with small holdings; indeed, a quarter of all the egg-laying fowls in the country are estimated to be on non-agricultural holdings and so escape enumeration in the census. Much poultry-keeping is associated with towns or with areas of small holdings or of part-time holdings, as in Cornwall. The importance of this consideration is, however, declining with the adoption of more intensive systems which favour large units; thus 18% of all adult fowls were on holdings with 500 or more birds in 1939 and 82% in 1970 with 26% in flocks of 20,000 or more.

The concentration of poultry in Lancashire is related to the size and kind of holding, but it also owes something to chance. Interest in poultry-keeping probably goes back to the early industrial period when workers often had small holdings. It was reinforced after the First World War when many unemployed industrial workers took up poultry-keeping. The system of housing birds favoured flat land and hence the coastal plain; but many concentrations here are associated with early pioneers of poultry breeding whose work firmly established Lancashire as the leading county for poultry and whose success further encouraged the concentration of the industry there.

Turkeys have long been associated with Norfolk and here, too, there is some tendency towards more intensive production in larger units. The predominance of Norfolk in duck-keeping is associated partly with the decline of duck egg production elsewhere, following a fall in demand, and partly with the rise of large units in which ducks are reared for meat; one farmer in Breckland was reported in 1962 to be producing 300,000 birds a year. Geese, which require little attention and feed largely on grass, are kept in small numbers on many family farms.

Numbers of poultry have varied greatly over the period since 1938, when there were 56,274,088. During the Second World War, these were reduced to a minimum of 29,121,127 in 1943, but numbers have since risen fairly steadily and by 1970 were 107% above those in 1938. Lancashire and the West Riding were the principal counties for poultry in 1939, but although they have remained important, their share of the total number of poultry in 1970 was 11%, compared with 17% in 1939. Eastern and southern counties, with the exception of Holland, had a larger share, and in Norfolk, which replaced Lancashire as the most important county, numbers increased by about 400%.

Numbers of the different classes of poultry also changed markedly. In 1938 there were 2,337,471 ducks, 608,715 geese and 788,515 turkeys, or respectively 4·2%, 1·1% and 1·4% of all poultry. The number of ducks fell to 1,611,552 in 1941, and after rising again to 2,521,122 in 1949, fell again and were 57% lower in 1958 than in 1938; the keeping of ducks has become increasingly concentrated in Norfolk, which had 40% of the ducks in 1970, compared with 16% in 1939. The number of geese declined little during the Second World War, but after rising to a maximum of 937,955 in 1949, has since declined steadily and was only 25% of the 1938 figure in 1970. The number of turkeys fell to 294,358 in 1943, but has risen since, particularly since 1955, and in 1970 was six times the number in 1938. Norfolk increased greatly in importance, with 26% of a much larger turkey flock in 1970, compared with 10% in 1938; the other major change has been the rise in Yorkshire's share of the turkey

Fig. 209. Geese per 1,000 acres of agricultural land.

Fig. 210. Number of livestock in livestock combinations.

flock, which increased from 5% to 21%. The major change in the number of fowls has been the increase in the number of broilers, output of which increased fifteen-fold between 1953 and 1958. They numbered 12,959,980 in 1960 or 15% of all poultry compared with 41,650,899 in 1970, or 36% of all poultry; yet even in respect of other fowl, Lancashire's share has declined from 12% to 9%.

Livestock Combinations

Figures 211 to 214 are an attempt to show the ways in which the different classes of livestock are associated in the various districts, using the method already employed for the identification of crop combinations (see Appendix I). In contrast to the maps of

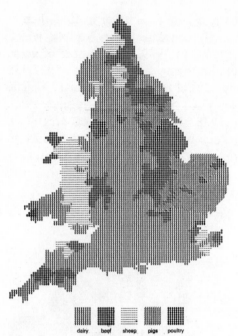

Fig. 211. Leading livestock in livestock combinations.

Fig. 212. Second-ranking livestock in livestock combinations.

Fig. 213. Third-ranking livestock in livestock combinations.

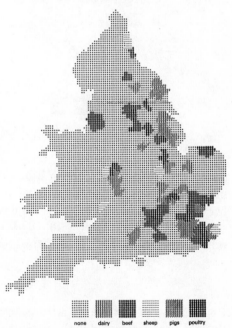

Fig. 214. Fourth-ranking livestock in livestock combinations.

enterprise and crop combinations, the maps of livestock combinations are relatively simple and, although this simplicity is partly due to the fact that sheep, cattle and other livestock are not subdivided into those kept for different purposes, e.g., breeding, rearing and fattening, it is largely the result of the dominant position of cattle.

Figure 210, recording the number of classes of livestock entering into livestock combinations, divides England and Wales broadly into two, though the division is less clear cut than in the case of cropping. Eastern England, with the exception of the Fenland and Sussex, has generally comparatively complex combinations, while those of western England and Wales are simpler, notably in the principal dairying areas; if the uplands did not lie in districts that included adjacent lowlands, many of them also would show only one class of livestock. Dairy cattle are the leading livestock in nearly two-thirds of the districts, the principal exceptions being the uplands, where they are replaced by sheep, and eastern England, where beef cattle or pigs are generally more important (Fig. 211). Beef cattle appear as the second-ranking livestock in over half the districts, though these are widely distributed, and dairy cattle in most districts where they were not the leading livestock (Fig. 212); poultry, too, appear as a significant element in Lancashire and in many parts of eastern and southern England. Third-ranking livestock are almost equally divided among the five classes, though sheep rarely occur in districts in eastern England, and a third of the districts have no third-ranking livestock (Fig. 213). Only a sixth of the districts, mainly in eastern England, have fourth-ranking livestock (Fig. 214); they are again almost equally divided among the different classes.

As the maps in the next chapter show, the space devoted to livestock in this atlas compared with that devoted to other aspects of the agriculture of England and Wales, is no measure of their great importance in the agricultural economy of the country. It is simply a reflection of the less satisfactory data which are available for the preparation of maps to show the regional character of livestock farming.

CHAPTER VIII

Enterprises and Farm Types

The maps in the preceding chapters show the distribution and relative importance of individual crops and classes of livestock. Yet these do not occur in isolation, but form parts of farming enterprises, e.g., dairying, poultry-keeping, the number and character of which define the type of farming practised. Individual farms can likewise be grouped in type of farming areas, throughout which similar types of farm occur. In general, the simplest farming systems are to be found where physical conditions for farming are most difficult and only a limited number of enterprises is possible; indeed, on a hill sheep farm there may even be a single enterprise, with sheep, supported mainly on rough grazings, the only livestock kept. On most lowland farms, however, several enterprises are generally undertaken, although some will be more important than others; thus dairying might be the principal activity, but pigs and poultry would also be kept and some crops sold. On the hill sheep farm, sheep are the most important enterprise by any criterion; all the land is used for their support, all the farm labour is devoted to them, and they provide, as wool, store lambs and draft ewes, the whole of the farm income. On most lowland farms, with a variety of enterprises, the identification of farming type is more difficult; for there are not, in fact, clear-cut types, but a continuum in which the different enterprises are present in many different permutations. Furthermore, somewhat different results will emerge if different criteria (income, labour use or land use) are employed. Thus even the most complex system of classification is a considerable simplification of reality; the same is true of type of farming maps, for in most areas it is rare for all farms to be of exactly the same kind.

In the absence of data for acreages and man-days on individual holdings, only rather general indications can be given of the distribution of different types of farm. It is possible to map the relative importance of individual enterprises, although it must be remembered that, while in some districts the enterprises recognized may well be typical of those on the great majority of farms, in others the enterprises may be only the average of two or more contrasting types. It is also possible to take the analysis a stage further by mapping the proportions of farms of different kinds (as classified by the Ministry of Agriculture), although such maps, too, should be interpreted with caution, not only because of the heterogeneous nature of the districts, but also because each farm is given the same weight irrespective of its size.

Man-days

Before the importance of different enterprises can be considered, they must be identified; this can be done on the basis of land use, labour requirements or sales of farm products. In view of the importance of livestock on British farms, it is difficult to employ land use alone as the basis of identification, but sales of farm products or labour requirements are possible criteria. In the absence of data about actual sales or labour used, it is necessary to convert the crop acreages and livestock numbers recorded in the agricultural census into money values or labour requirements, using standard factors derived from the study of many individual farms. Standard outputs, in which money values are attributed to the acreages of crops sold and to the numbers of livestock, have been used in classifications of farms in Dorset and East Anglia, while standard labour requirements, which have the advantage of changing fairly slowly, have been favoured in official classifications and are employed here. Equally valid, though somewhat different maps might have been prepared by using standard outputs.

Standard labour requirements are clearly open to many objections. Partly owing to the differences noted earlier in the degree of mechanization, in field and farm size, and in terrain, it is unlikely that labour requirements for the same crop are everywhere identical. Yet, the use of different factors in each region raises as many problems as it solves, while the fact that western and northern counties are predominantly pastoral, and eastern counties mainly arable minimizes the importance for man-day calculations of regional differences in crop and livestock labour requirements. In any case, such differences as exist are more likely to affect the absolute value of labour requirements than the relative importance of different activities. Whatever its limitations, the approach has been thought sufficiently reliable to be used in distinguishing farms eligible for help under a statutory scheme for aid to small farmers.

The factors selected are those used by agricultural economists in the Ministry of Agriculture and are set out in Appendix I. Each acreage and the number of each class of livestock had to be multiplied by the appropriate factor and the results added to give the total man-days for each district. Man-days for the individual crops and livestock had then to be allocated to the appropriate enterprise in preparation for computing both the relative importance of each enterprise and enterprise combinations, using the method already described for identifying crop and livestock combinations (Appendix II). Apart from these calculations, the number of man-days for every 100 acres of agricultural land was also computed after the addition of 15% to cover the labour requirements of overheads.

Results of these last calculations are plotted in Figure 215 and give some indication of the intensiveness of farming. The requirements are clearly related to farming type, although, for reasons noted above, this measure of the absolute level of labour requirements is probably less reliable than the assessment of the relative importance of different enterprises recorded in Figures 216–222. Areas where labour requirements

207

Fig. 215. Man-days per 100 acres of agricultural land.

are less than 300 man-days per 100 acres are confined to the uplands of Dartmoor and Exmoor, Wales, the Pennines, the Lake District and the North York Moors, where extensive sheep rearing is practised. In most of Wales and northern England labour requirements are fewer than 500 man-days, the principal exceptions being west Cornwall and the Tamar and Exe valleys. The remainder of the lowlands, i.e., those where dairying and arable farming are practised, require between 700 and 1,000 man-days. Districts recording even higher values of 1,000 man-days and over are chiefly those where intensive horticulture is practised, such as the Fenland, Worcestershire, Bedfordshire, Kent and other parts of the Home Counties; the Fylde is the main exception. The highest value recorded is 1,873 per 100 acres in District 155 and the lowest 60 in District 34.

Enterprises

The allocation of man-days to different enterprises presented greater problems. Initially it was decided that crops and livestock should be grouped in seven enterprises, viz., cropping, dairying, beef cattle rearing and fattening, sheep rearing and fattening, horticulture, pigs and poultry. The chief difficulties arose over the allocations of the man-days due to crops and to grass and over the distinction between beef and dairy cattle; both problems are important because cash cropping, dairying and rearing and fattening beef cattle are all important enterprises and show markedly different regional preferences. Minor difficulties were also experienced over the allocation of horticultural crops. In the absence of data about crop disposal in each district it was necessary to decide whether a distinction could be made between cash crops and other crops. Some crops are undoubtedly grown primarily for sale, e.g., wheat, sugar beet and potatoes, but others, notably barley, are both cash crops and consumed in large quantities on the farms on which they are grown; still others, such as kale, are used entirely on the farm. An arbitrary decision could be made to allocate a fixed proportion of crops such as barley to cash cropping, though true proportions are likely to vary considerably throughout England and Wales. Alternatively, all crops could be allocated to an enterprise 'cropping', and, while this would diminish the relative importance of those areas where crops are grown primarily for sale, it was decided to adopt the latter solution. A further aspect of this problem is that some vegetables are grown on a field scale on arable farms and so are more properly allocated to cropping than to horticulture. The Ministry of Agriculture has devised rules based on the number of crops grown, but these can be applied only to data for individual farms. It was therefore decided to allocate broad and French beans grown for processing and peas for processing and for harvesting dry to cropping and the remaining vegetables to horticulture.

Grass presents a different problem, for it is all used for the support of livestock. Consumption by pigs and poultry forms a negligible proportion so that some way had to be devised to allocate man-days due to grass among dairy cattle, beef cattle and sheep. Ministry practice was followed in allocating grass man-days according to the respective requirements of starch equivalent from grass of the different classes of grazing stock, estimated at 1·56 per man-day of dairy cattle, 3·9 per man-day of beef cattle and 12·0 per man-day of sheep. The man-days attributed to each class of stock were then multiplied by these factors and the grass man-days allocated between them in proportion to their share of all starch equivalent.

The third major problem was the distinction between beef and dairy cattle, a division that is inevitably somewhat arbitrary in view of dual purpose breeds and of the large proportion of the cattle slaughtered for meat which is derived from the dairy herd. In the 1970 census only cows and heifers in milk and in calf were divided into those primarily kept for dairying and those kept primarily for beef production. The basis of allocation of 'other female cattle' between beef and dairy enterprises was the relationship between the number of such cattle (subdivided into those 1 year old and over and younger cattle) and the number of first-line replacements (heifers in calf with first calf). If the latter was greater than or equal to the number of other female cattle 1 year old and over, these were all allocated to dairying; if it was smaller, only the equivalent number of other cattle was allocated to dairying and the rest to beef. A similar procedure was followed in respect of female cattle under 1 year old. Such a division is, of course, a rough approximation, for the ratio of dairy cows to herd replacements varies throughout England and Wales and some of the cattle treated as replacements are, in fact, culled cows to be slaughtered for beef. All male cattle other than bulls and bull calves were allocated to the beef enterprise and bulls and bull calves were divided according to the ratio of dairy cows and heifers to beef cows and heifers.

When man-days had been allocated to the various enterprises according to these rules, the proportion of total man-days which was attributed to each enterprise was calculated; some of the results are shown in Figures 216–222. These maps show only the relative importance of different enterprises; their absolute importance must be judged from the maps in the preceding chapters. In Figures 216–218, man-days allocated to dairying, beef and sheep include the appropriate allocation of grass man-days. Dairy cattle, the leading enterprise on a large number of farms, account for 60% or more of all man-days in the main dairying areas: Cheshire, Derbyshire and Staffordshire; south-west Wales; and Dorset and Somerset (Fig. 216). The highest proportion is 78% in District 81. Proportions are also high in the southern Pennines and north-west England, and lowest in eastern England (especially the Fenland) and parts of the uplands. Beef cattle are nowhere as important, the highest proportion being 38% in District 3. They account for 20% or more in three main areas, viz., north-east England, upland Wales and south-west England; they are least important, in summer at least, in the main arable areas, especially those where horticultural crops are common, but account for 5–10% over most of the lowlands, with highest values in the east Midlands (Fig. 217). Sheep are important in much the same areas, though the south-west peninsula is less prominent, with only Exmoor having a proportion of 20%

Fig. 216. Dairy cattle man-days per 100 man-days.

Fig. 217. Beef cattle man-days per 100 man-days.

Fig. 218. Sheep man-days per 100 man-days.

Fig. 219. Crop man-days per 100 man-days.

211

or more; upland Wales has large areas with 40% or more, reaching 68% in District 316A (Fig. 218). Proportions in the lowlands are generally under 5% except in the east Midlands and Kent.

The map showing the importance of crop enterprises reveals a very different pattern, with the highest values (of 40% and over) in eastern counties, especially those from Suffolk to the East Riding (Fig. 219). Only Holderness, the Fens and Broadland are anomalous. South-east England is, by contrast, an area of comparatively low values. Further west, values are generally below average, except in the chalklands, the west Midlands, south-west Lancashire and Pembrokeshire. Outside eastern England, this map may also be interpreted to give some indication of the relative importance of livestock by using the complement of the percentages shown; most western districts record more than 80% of their man-days from livestock enterprises.

The patterns shown in Figures 220–222 are far less clearly defined, a reflection of the fact that, even at this level of generalization, these enterprises are much more localized; all, of course, show lowland distributions. Yet, while pigs are relatively important in the lowlands, only in East Anglia and in Holderness and around the London and south Pennine conurbations do proportions exceed 10%, the highest value being 27% in District 32 (Fig. 220). For poultry, the pattern is even more highly fragmented, south Lancashire and the West Riding providing the only large block where poultry account for 15% or more of all man-days, though significantly in view of the regional changes in the distribution of poultry, the highest value (47% in District 169) is in Norfolk (Fig. 221). Horticultural enterprises also show a more fragmented pattern than do crop enterprises, although the pre-eminence of east and south-east England is clearly marked, with the Fens and districts on either side of the Thames Estuary as the most important areas (Fig. 222). There are several outliers, with Worcester as the most important; others include Bedfordshire, south Hampshire, south-west Lancashire, the West Riding, the Humber warplands, south Devon and west Cornwall.

Enterprise Combinations

Enterprise combinations have been computed in the same way as crop and livestock combinations; the results are given in Appendix III. Seven enterprises are considered, viz., those mapped in Figures 216–222, and the number occurring in the combination for each district is shown in Figure 223, which can be interpreted as an indicator of variety in farming systems. In a broad view, it shows a similar three-fold division to that identified on the map of crop combinations (Fig. 98), although the central belt now has the more complex combinations and the flanking belts the simpler.

Figures 224–227 show the first four ranking enterprises in these combinations, though the omission of an enterprise from these maps or from the combinations listed in Appendix III does not imply that it is absent from that district, but merely that it is not sufficiently important to appear in that combination. As on the similar maps of

Fig. 220. Pig man-days per 100 man-days.

Fig. 221. Poultry man-days per 100 man-days.

Fig. 222. Horticultural man-days per 100 man-days.

Fig. 223. Number of enterprises in enterprise combinations.

213

crop and livestock combinations, a dot is used to identify those districts in which no enterprise occurs in the combination at that level, and such dots consequently occupy a progressively larger part of the map with each successively lower-ranking enterprise. Dairying is the leading enterprise in nearly half the districts, including nearly all the south-west peninsula and most of north-west England and lowland Wales (Fig. 224); cropping accounts for most of the remaining districts, with sheep in the uplands, horticulture in south-east England and other areas of intensive horticulture, and beef cattle and poultry in a few districts. The map of second-ranking enterprises shows beef cattle succeeding sheep in most of the uplands, dairying in south-west England, dairying or beef cattle following crops in the east Midlands, and horticulture following cropping over much of eastern England (Fig. 225). There is no third-ranking enterprise in nearly half the districts, which are located mainly in the uplands, or in the principal dairying areas (where dairying is generally sufficiently important to be the only significant enterprise), or in the main areas of cash crops production (Fig. 226). Beef cattle are the most important enterprise at this level, occurring widely throughout the Midlands and the upland margins; East Anglia is anomalous among the areas of arable farming in that poultry or dairy cattle appear as third-ranking enterprises. Three-quarters of the districts have no fourth-ranking enterprise; the remainder are located mainly in the central belt already identified and each of the seven enterprises makes a contribution (Fig. 227).

In interpreting these maps it must be remembered that they represent enterprises by districts, i.e., they describe the importance of enterprises on each 'district farm'. Some of these enterprises will be concentrated on specialist holdings, while others will occur as main or subsidiary enterprises on generally large farms. The run of the district boundaries, particularly in the uplands, may also blur regional contrasts which are quite clearly defined on the ground. Thus, the uplands of the Lake District are hardly represented on these maps because most districts in Cumberland include both upland and lowland, and the distinctiveness of the Cotswolds and of the limestone escarpments generally is minimized by the diversity of terrain included in the districts in which they lie. The use of a measure of intensiveness, viz., man-days, as the basis for identifying enterprise combinations may also lead to some apparent anomalies; for an intensive enterprise, such as poultry-keeping in District 169, may result in a district being identified with an enterprise which occupies only a very small part of the area. The identification of dairying as the leading enterprise in each district in Wiltshire may similarly seem at variance with the visual impression that all the chalk uplands are devoted to cropping, but it is nevertheless supported by other evidence.

Farm Types

Since 1963, the Ministry of Agriculture has classified the holdings enumerated in the agricultural census, distinguishing part-time from full-time holdings and subdividing the latter into types of farm according to the number of man-days devoted to the

Fig. 224. *Leading enterprise in enterprise combinations.*

Fig. 225. *Second-ranking enterprise in enterprise combinations.*

Fig. 226. *Third-ranking enterprise in enterprise combinations.*

Fig. 227. *Fourth-ranking enterprise in enterprise combinations.*

different enterprises. The results have been published in a series of studies, the latest of which, *Farm Classification in England and Wales 1969–1970*, includes analyses of the 1970 census data. The Ministry has also published maps of farming types, based on a sample of holdings and using the 10 kilometre squares of the National Grid as mapping units. Only numbers of holdings are recorded in the district summaries and Figures 228–233 show the proportion of full-time holdings in each of the main types.

Full-time holdings are divided under the Ministry's classification scheme into thirteen types, the dividing line between full-time and part-time holdings being 275 man-days. These types are grouped into six main types and it is these which are mapped; dairying, livestock, cropping, pigs and poultry, horticulture and mixed. In all but the last, the type of farm is identified by more than 50% of the man-days being in one enterprise, the mixed class consisting of those holdings (10% of the total) where no one enterprise predominates. The table below gives the proportion of all full-time holdings of various types and of each type by regions.

Region / Type	Eastern	South eastern	East Midland	West Midland	South western	Yorks and Lancs	Northern	Wales	England and Wales
% full-time holdings in each region									
Dairy	6	30	31	49	56	40	40	47	38
Livestock	2	10	10	15	16	8	28	42	17
Cropping	50	15	36	11	5	2	13	1	18
Horticulture	24	23	5	9	5	11	2	2	10
Pigs and Poultry	11	12	7	6	6	12	4	2	7
Mixed	7	10	10	10	12	8	13	7	10
All f.t. holdings	100	100	100	100	100	100	100	100	100
% full-time holdings of each type									
Dairy	2	8	8	16	28	12	10	16	100
Livestock	1	6	6	12	18	5	17	34	100
Cropping	38	9	19	8	5	14	7	1	100
Horticulture	33	24	5	11	10	12	2	2	100
Pigs and Poultry	21	17	10	10	15	18	6	3	100
Mixed	11	10	10	13	23	10	13	10	100
All f.t. holdings	14	10	10	13	19	11	10	13	100

These values reveal the pre-eminence of dairy farms and also the degree of regional specialization; mixed farms are the only type to be fairly evenly distributed among the different regions.

Figures 228–233 add a further dimension to this analysis by recording the proportion of full-time holdings of each type by districts. Dairy farms predominate, both by area and by number of holdings, in three main areas, namely the principal dairying areas already identified, at the cores of which 70% or more of full-time holdings are dairy farms; proportions are lowest in upland Wales and the main areas

216

Fig. 228. Dairy farms as a percentage of all full-time farms.

Fig. 229. Livestock farms as a percentage of all full-time farms.

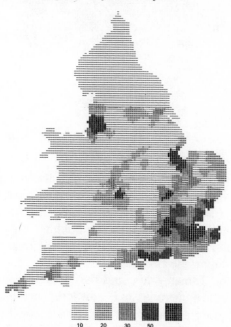

Fig. 230. Cropping farms as a percentage of all full-time farms.

Fig. 231. Horticultural farms as a percentage of all full-time farms.

217

of horticultural and cash crop production (Fig. 228). Livestock farms embrace a wider variety of types, from the hill sheep farms of upland Wales to the cattle rearing farms of the Welsh border and the fattening farms of the east Midlands (Fig. 229). Proportions are highest in Wales, exceeding 50% over most of the uplands, and 90% in Radnorshire; the lowest proportions are also in the main arable areas, though only in the east Midlands are there above-average values in the lowlands. Cropping farms are located mainly in eastern England, accounting for half or more of the farms in a continuous belt from the Yorkshire Wolds to north Essex, but with lower values in the Fens, the Lincoln marshes and Broadland (Fig. 230); there are other outlying areas

Fig. 232. Pig and poultry farms as a percentage of all full-time farms.

Fig. 233. Mixed farms as a percentage of all full-time farms.

where cropping farms are relatively important, on the chalk and limestone uplands and in the west Midlands, but in most western districts the proportion is under 10%. Horticultural holdings are much more localized and less numerous, and account for half or more of full-time holdings only in the Fens, Bedfordshire, Worcestershire and north Kent (Fig. 231); other areas where horticultural holdings are relatively important are south-west Lancashire, south Hampshire and Sussex, the East Riding, the Tamar valley and the Home Counties generally. Pig and poultry farms are also less numerous and nowhere account for half the holdings; they are relatively important in south Lancashire and the Pennines, East Anglia, Hampshire and the Home Counties (Fig. 232). Mixed farms are widespread, but are relatively most important in north-west England, the Midlands and south-west England, and least in

218

the main dairying areas and the Fenland (Fig. 233). Mapping the proportions of farms of different kinds gives one indication of their distribution and relative importance, but the pictures these maps convey correspond to a considerable degree with those shown by unpublished maps of their contribution by acreage and by the maps of enterprises already discussed. Nevertheless, it must be remembered that they are very generalized and convey only a subdued impression of the rich variety of agriculture in England and Wales.

CHAPTER IX

Retrospect and Prospect

The ideal agricultural atlas would face no limitations imposed by time, cost or inadequate data; but, in practice, the preparation of any atlas involves compromise and choice. The purpose of this chapter is to indicate some of the steps which are necessary for the production of a more adequate atlas at some future date.

Two of the major limitations of the present atlas, the relatively coarse grain of the picture provided by the district data and the essentially static view of the agricultural scene which it presents, could have been met, at least in part, if greater resources had been available, though there would still have been practical problems to overcome. It would clearly have been preferable to produce maps, not only of mean values for several years, but also of change, both short- and long-term; but this would have resulted in a much longer and very different book; furthermore, comparisons are made difficult by changes in the items recorded in the census and in the boundaries of districts and parishes. Maps of much finer grain could have been made by using parish data, but such maps would not only have required more computer time, but might well have been more difficult to interpret at this scale, as Figures 15–16 suggest; some degree of generalization is clearly necessary if such complex data are to be grasped. A compromise solution might be to group parishes into much more uniform statistical units than the districts, e.g., areas approximating to 10 kilometre squares of the National Grid; but the practical difficulties of doing this are considerable and it was not possible within the time-scale available for this revision. Some of the practical difficulties which handicapped preparation of the first edition have since been overcome. The making of maps automatically is now practicable, though there is undoubtedly considerable scope for improvement; we are still using devices designed primarily for printing words or for simulating the activities of a draughtsman. However, it is now quite feasible to produce an annual atlas of agriculture as readily as annual volumes of statistics are produced by census offices. Inability to handle large quantities of data, a major restraint on the production of agricultural atlases in the past, is now largely removed with the advent of large digital computers and mapping programs, though there are still considerable organizational problems in handling large bodies of data efficiently; what remains are the limitations of the data to be mapped and the absence of data on many aspects of agriculture which are essential to an adequate understanding of the many major and minor differences in the character of agriculture throughout England and Wales.

The annual agricultural censuses are not, of course, undertaken with the aim of analysing such regional variations, and modifications to the census to meet this need can hardly be expected, though it is interesting to speculate on the character of the data which ought to be collected if cost and the ability and willingness of farmers to supply information did not pose limitations on the form of the census and if its collection were to be organized today for the first time and not with the inertia and precedent of more than a hundred years of census-making. Unfortunately, some developments in methods of collecting agricultural data are tending to make the material less rather than more useful to geographers and map makers, for there is an understandable tendency for data (and often the more interesting data) to be collected from sample farms so that sampling errors severely limit the extent to which such data can be mapped, even with the county as the basic statistical unit. Nevertheless, there are feasible improvements that might be beneficial. It is reasonable to suppose that full-time farmers at least are now sufficiently practised in keeping records for questions to be asked relating to a breeding or cropping season as a whole, e.g., the number of lambs born on a holding during a lambing season, and not merely to the situation on the day of the census; indeed, a start has already been made with questions about the number of calvings on the holding in the preceding quarter. There is also a strong case on this and other grounds for replacing the present annual census by an exhaustive quinquennial census, in which more sophisticated questions could be asked and, if necessary, answered by field enquiry, with only summary censuses confined to the major crops and livestock in the intervening years. At the least, it ought to be possible, now that most records are kept in machine-readable form, to bring together some of the diverse information on other inputs and on the disposal of farm produce that are now collected in other ways for a variety of purposes. Perhaps, too, a start could be made on collecting the co-ordinates of the boundaries of each holding or, at least, of some reference point within it and any detached portions; such information would facilitate not only the administrative task of monitoring changes in farm size and the extent of farm land, but also the construction of more significant maps.

A major omission in the mappable data available concerns qualitative aspects of the country's agriculture of a kind not generally recorded in agricultural censuses, e.g., regional differences in breeds of livestock and varieties of crops, in systems of livestock husbandry and crop rotations, and in the timing of the seasonal round of activities which make up the farming year. Much of the information already exists in the minds of those whose lives are spent in agricultural administration, advisory work or research; the collective knowledge of the staff of the Ministry of Agriculture alone must be immense. Thus, if the intimate local knowledge of district and specialist officers of the Agricultural Development and Advisory Service could be collected in a way which would permit comparison from area to area, e.g., by a systematic questionnaire, many of these gaps could be filled. Research institutes, university departments, large commercial organizations and the farmers' unions are similarly repositories of vast stores of unrecorded knowledge, all of which could be exploited without breach of confidence if there was an awareness of its interest and value.

A deficiency which could not so easily be made good is the lack of information about the ways in which crops and livestock move from farm to farm and from farm to market. It would be impracticable to think of acquiring or representing data for all such movements throughout the country, but it is highly desirable that sample studies should be made to determine their scale and direction, particularly for sheep and cattle, whose movements are probably the most complex of all. Investigations could profitably be undertaken of the range of movement in sample areas and of the countrywide movement of selected classes of stock: it would also be instructive to follow the movement of individual animals throughout their lives. Such studies could most easily be undertaken where there is some form of central control, as with the marketing of milk. It seems likely that records of the veterinary services might be used for some of these purposes and W. J. Carlyle has made extensive use of the records of auctioneers in his study of the movement of store stock in Scotland.

The lack of knowledge, at least on paper, of the reasons for regional differences in crop and livestock farming is a challenge to workers in many disciplines. The answers can be sought by two quite different approaches: through the investigation of the origins of regional specialization and through the analysis of the physical, social and economic advantages which various parts of the country enjoy. These two approaches are, of course, complementary, for once origins have been traced, they must still be related to physical and other advantages, while past commitments in terms of buildings, farm layout, farm equipment and farming skills, act as a brake on changes in farming systems.

At present, for want of a better term and for lack of precise knowledge, regional specialization in a particular crop or class of livestock is often explained by the assertion that it is traditional, implying merely that it has been long established. It is, indeed, surprising how persistent are many of the major features of the agricultural geography of the country, yet the origins of few have been traced; the development of the present pattern of liquid milk production is one of the few major topics to have been treated in this way. The origins of many features are so remote that investigations are often handicapped by lack of data; but for most crops and livestock it is possible to go back at least to 1866, when the first regular collection of agricultural statistics began, and generally to the Board of Agriculture Reports *c.* 1800. The localization of many of the minor crops is of more recent origin; horticultural cropping provides many examples which are worth investigation.

It is, of course, important to distinguish between the introduction of a crop or animal into a particular locality, which is often difficult to trace, and its emergence as a major regional characteristic; thus the growing of vegetables around Sandy long preceded the rise of the mid-Bedfordshire market gardening region. Even when origins cannot be traced, it is interesting and illuminating to investigate the degree of stability, in both the short and the long term, of crop and livestock distributions, particularly in view of the lack of suitable economic data. Studies of post-war changes have been made in respect of dairy cattle, barley and potatoes, but there are many interesting regional changes which await proper analysis; examples are the extension of the area under kale, the contraction of that under oats and mixed corn, the numerous changes in the pattern of fruit and vegetable growing, the movement out of milk production, the expansion of beef production and the rise of the broiler industry. Some of these topics have, of course, been examined from a national point of view, but their regional aspects have generally been neglected.

While it is always interesting to know when and how regional specializations arose, they can also be explained by reference to comparative advantage, which is generally inferred from differences in physical geography. Yet the reality of these alleged physical or other advantages has often been accepted without question. The work of the Soil Survey will gradually provide a means of testing assumptions about soil differences; meanwhile, it would be useful to have investigations of sample sites to see whether there is *prima facie* evidence of meaningful physical differences, e.g., between the soils of the Market Harborough area which support high grade fattening pastures and those of other parts of the east Midlands which do not. There is, similarly, scope for the investigation of other alleged advantages, e.g., whether Kent, as the premier fruit county, does enjoy any marked climatic superiority over other areas in south-east England, or whether proximity to urban markets does, in fact, confer any benefits in vegetable growing.

Yet, while it is necessary to establish the reality or otherwise of such advantages, comparative advantage ought not to be inferred indirectly from them, but measured independently in economic terms. There are obvious difficulties, particularly the great range of inter-farm differences, but a vast body of material has been and is being collected by agricultural economists in universities, the Ministry of Agriculture and elsewhere about the profitability of different enterprises. Unfortunately, for present purposes, many of the results of these investigations are given as averages for whole provinces or in terms which make it impossible to compare differences between provinces. While samples may be too small for detailed regional analysis, presentation of results for named localities would give some idea of intra-provincial differences which can otherwise only be assumed. What is needed is information, on a uniform basis, about the costs of and returns from the same enterprise in different parts of the country and different enterprises in the same locality, similar to that produced for the national investigation into the economics of milk production. The 2,500 or so farms of the Farm Management Survey ought to be a sufficiently large sample to yield valid statements about costs and returns for different enterprises in all the major agricultural regions of the country. The development of the gross margin technique should also result in the production of data which can clarify the economic logic of the very real degree of regional specialization which has been shown to exist.

By adopting techniques for the rapid compilation of maps, by improving the range and quality of mappable data and by investigating the economic, physical and historical bases of particular crop and livestock distributions, it should be possible to prepare a much more representative collection of maps whose explanation would depend less on unsupported assertion and more on careful investigation. Such maps would have not only educational value; they could aid the monitoring of the effects of policies and even policy-formulation itself. In no other way can the diversity of both natural and human resources be so readily comprehended by the policy-maker and in no other way can he so adequately appreciate the varying reactions throughout England and Wales to the policies he devises.

Sources and Methods

Choropleth Maps

Class intervals on choropleth maps have been chosen to show the range of values in the country as a whole and have been selected after an inspection of frequency graphs, estimates of means and standard deviations, tabulated results, and trial and error. As far as possible equal intervals have been used and numbers of classes have been restricted to six on full-page maps and five on those occupying a quarter page. On maps showing highly localized distributions geometrical scales have been used and some crops are mapped as a proportion of the total acreage in England and Wales, rather than as densities per 100 or per 1,000 acres. Care is therefore needed in interpreting the various maps, for while the shading will immediately indicate which areas are relatively most important and which least, they will themselves give no indication of the values represented by each class. Figures 17 and 19 provide a striking example of the danger of superficial comparison, for although both sets of class intervals increase by equal steps, the base of the top class is twelve times greater in one map than in the other. On maps produced by line-printer, the value shown underneath each block represents the upper limit of that class (or the maximum if no number is shown); the lower limit is the maximum of the class below (or 0 if there is no lower class).

Sources and notes on the construction of individual maps

Figure 1. Based on maps of the Ordnance Survey.

Figure 2. Constructed by superimposing the 10 kilometre National Grid on the 1 inch to 1 mile maps of the Land Utilisation Survey, and establishing for each grid square the maximum height of the boundary between continuous moorland and cultivated land.

Figures 3 and 4. Based on maps 13 and 18, *Climatological Atlas of the British Isles*, H.M.S.O., 1952, 72 and 77.

Figure 5. Based on Map 4, Technical Bulletin No. 4, *The Calculation of Irrigation Need*, H.M.S.O., 1954.

Figure 6. Based on a map prepared by Dr. S. Gregory in *Trans. Inst. Brit. Geog.*, **20**, 1954, 65.

Figure 7. Based on map 7, *Climatological Atlas of the British Isles*, 125.

Figure 8. Based on a map by W. N. Hogg, in *The Biological Productivity of Britain*, Institute of Biology Symposium, London, 1958.

Figure 9. Based on a map prepared by Dr. D. A. Osmond, Soil Survey of England and Wales, for F.A.O.

Figure 10. Based on the Land Classification map prepared by the Land Utilisation Survey and published in L. D. Stamp, *The Land of Britain: its use and misuse*, 3rd Edition, 1962.

Figure 11. Based on maps of the Ordnance Survey.

Figure 12. Based on lists of parishes and districts supplied by the Agricultural Census and Surveys Branch, M.A.F.F.

Figures 13–14. Based on unpublished district summaries of the agricultural returns, 1970.

Figures 15–22. Based on unpublished holding data, by parishes, agricultural returns, 1970.

Figure 23. Based on data in *Technical Rept. 19/2*, Agricultural Land Service, M.A.F.F., 1971.

Figures 24–29. Based on unpublished district summaries of the agricultural returns, 1970.

Figures 30–41. Based on unpublished estimates, derived from one-third samples of all holdings at the September, December and March censuses, 1970–1.

Figure 42. Based on information supplied by the Milk Marketing Board.

Figure 43. Based on information supplied by the British Sugar Corporation.

Figure 44. Based on a map in *Report of the Reorganisation Committee for Eggs*, Cmnd. 3669, H.M.S.O.

Figure 45. Based on information supplied by the Fruit and Vegetable Canning and Quick Freezing Research Station.

Figures 46–233. Based on unpublished district summaries of the agricultural returns, 1970.

Figure 167. Livestock units have been identified by multiplying the numbers of various classes of livestock by the following conversion factors, as adapted from *Terms and Definitions used in Farm and Horticultural Management*, M.A.F.F., 1970: dairy cows, 1·0; heifers in calf (first calf), other male cattle 2 years old and over, 0·80; bulls, other female cattle 1 year old and under 2, 0·70; beef cows, other male cattle 1 year old and under 2, 0·60; bull calves, other cattle under 1, 0·40; sheep 1 year and over, 0·10; lambs, 0·05; breeding sows, gilts in pig, 0·50; boars, 0·40; other pigs 2 months old and over, laying hens, fowls for breeding, turkey hens for breeding, 0·02; pullets to point of lay, table fowl other than broilers, ducks, geese, other turkeys, 0·005; broilers, 0·002.

Figure 215. Man-days have been calculated using the following factors:

Field Crops	Man-days per acre		Man-days per acre
Wheat, barley	2	Beet, lettuce	30
Oats, mixed corn, maize	3	Broad beans and green peas (market)	30
Beans for stock, oil seed rape	3		
Potatoes	15	Salad onions	100
Sugar-beet	10	Runner beans	45
Hops	70	Runner beans, climbing	80
Turnips, swedes, fodder-beet	9	French beans, market	35
Mangolds	11	Broad, French beans and green peas (processing)	4
Rape for stock	1		
Kale for stock	1·5	Celery	40
Cabbage, savoy, kohl rabi for stock	5	Other vegetables	50
		Grass	1,300
Mustard, other crops	3		
Bare fallow	0·5		Man-days per head
Lucerne	1·25	Livestock	
Temporary grass	0·75	Dairy cows	10
Permanent grass	0·5	Beef cows	3
Vegetables		Dairy heifers in calf	3·5
Brassicas, turnips and swedes	20	Beef heifers in calf	2·5
Carrots, early	60	Bulls	6
Carrots, maincrop and processing	10	Other cattle	2·5
Parsnips, dry onions	25	Ewes and rams	0·7

	Man-days per head		Man-days per acre
Wethers and other sheep 1 year old and over	0·2	Cider apples, perry pears	8
Lambs under 1 year old	0·0	Cooking apples, plums, other top fruit	20
Sows for breeding, boars	4	Non-commercial orchards	1
Barren sows and other pigs 2 months old and over	4	Nursery stock, fruit and ornamental	50
Other pigs under 2 months old	0·0	Roses	100
Ducks	0·2	Other nursery stock	200
Laying hens, geese, turkeys, pullets	0·05	Strawberries grown under cloches etc.	150
Broilers, other table birds	0·05	Strawberries grown in open	70
Fowls for breeding	0·15	Raspberries	80
		Blackcurrants for market	45
Other horticultural crops	Man-days per acre	Blackcurrants (processing), gooseberries	40
Dessert apples, pears	25	Other small fruit	60
Cherries	30	Flowers	180

Figures 228–233. Based on classifications of all full-time holdings recorded in the June 1970 census; the method of classification is outlined in *Farm Classification in England and Wales 1969–1970*, M.A.F.F., H.M.S.O., 1971.

Crop, Livestock and Enterprise Combinations

The method used to establish crop, livestock and enterprise combinations is that devised by J. Weaver for his study 'Crop combination regions in the Middle West' (*Geogr. Rev.*, **44**, 1954, 175–200), as modified by D. Thomas (*Agriculture in Wales during the Napoleonic Wars*, 1963, 80–1). Since the method is the same in all these cases, its application will be illustrated by reference to crop combinations. It consists of comparing the actual percentage of the tillage area occupied by the different crops with theoretical distributions in which the tillage area is equally divided among the component crops. The purpose of the procedure is to group districts with similar cropping, by establishing which theoretical combination the actual distribution of crops most closely resembles.

The procedure is as follows. Crops are ranked in order of descending magnitude and the differences between the actual and theoretical percentages calculated, beginning with monoculture, in which one crop accounts for 100% and proceeding from 2-crop combination (each crop 50%) through progressively larger combinations to find the best fit. This is determined by the method of least squares. The differences between the actual and theoretical percentages are squared and summed, the lowest total identifying the desired combination. Thus in a district where the ranked percentages of the first ten crops are 33·7, 29·5, 14·6, 9·0, 3·7, 2·7, 2·1, 1·8, 1·0 and 0·6, the first calculation is $(100+3307)^2+(29·5-0)^2+(146-0)^2+(9·0-0)^2+(3·7+0)^2+(2·7-0)^2+$

227

$(2 \cdot 1 - 0)^2 + (1 \cdot 8 - 0)^2 + (1 \cdot 0 - 0)^2 + (0 \cdot 6 - 0)^2 = 5,594$. The second calculation is $(50 - 33 \cdot 7)^2 + (50 - 33 \cdot 7)^2$, and then as for the first calculation, giving a total of 1,013. The third calculation begins $(33 \cdot 7 - 33 \cdot 3)^2 + (33 \cdot 3 - 29 \cdot 5)^2$ etc., giving a total of 475, which is, in fact, the lowest, and hence identifies the cropping of this district as a 3-crop combination. The combination is described by appropriate initials for the component crops given in rank order, e.g., BOT, barley (first crop), oats (second crop) and turnips and swedes (third crop).

Mapping by Line-printer

J. McG. HOTSON, M.A.

Since the first edition of this atlas there has been a marked development in the use of computers in geographical research. A measure of this progress is that, while the more complex calculations for maps in the first edition were done by computer, the maps themselves were all hand-drawn, whereas nearly all the maps in this edition have been entirely produced by computer. Moreover, though the maps shown in this atlas are numerous, illustrating a wide sprectrum of agricultural activity, they form only a small fraction of the kinds of maps it is now possible to produce, for they are the product of only one output device, the line-printer, and are restricted to the relatively straightforward aspects of agricultural activity which are mappable from census data by choropleth maps.

The map work illustrated in this atlas, which is based on the CAMAP Programs developed in the Geography Department at Edinburgh University, can be conveniently considered from two view points, theoretical and practical. The theoretical aspect is concerned with the basic principles of this kind of computer mapping by line-printer, that symbols in a particular pattern can be used to give spatial meaning to tables of figures, whilst the practical aspect covers the problems encountered in actually producing maps.

The ability of computers to do large numbers of calculations on the data stored in data banks is well-known, for this is the main purpose for which computers were designed. The latest generation of computers can handle more data and do calculations more quickly and with greater efficiency than their predecessors, and rapid developments are in progress. Yet, impressive as these computational powers are, geographers, amongst others, are more interested in the spatial aspects of data and are learning to manipulate the output of computation to provide a spatial answer to a calculation instead of a single figure, or table of figures.

For some time, there has been a number of output devices to satisfy this demand, comprising plotters of different kinds. These work either with a large stationary flat bed or a rotating drum on which a piece of paper is fixed and over which a pen moves under the control of a computer. The maps produced in this way are similar to the traditional type of map which cartographers draw and with which geographers and others are familiar, and these are accordingly visually acceptable. This process of drawing maps is inevitably comparatively slow, even though the computerized calculations are quick, for much time is wasted in the mechanical movements of the plotters. For some time, then, geographers have turned their minds to the use which can be made of faster output devices which, although they may produce maps which are less satisfactory from the viewpoint of conventional cartography, can nevertheless produce them very quickly to illustrate adequately the spatial aspect of calculated results.

The line-printer has frequently been used, mostly because it is the common output device attached to all large computers, but the real possibilities of mapping by line-printer are only now coming to be fully appreciated. Probably its greatest advantage, other than speed, is that it operates on a fixed grid in a similar way to a typewriter, with sequential movements along the line from left to right, and down the page line by line. Spatial data can thus be fed to the printer in the order in which they are to appear on the map in exactly the same way, i.e., left to right across the map, grid-cell by grid-cell, and line-by-line down the map. For such mapping it is therefore unnecessary to attach co-ordinates to the data to enable them to be mapped as long as they are presented in the correct order and in the correct way. At least two-thirds of the data formerly required for mapping can therefore be omitted. Choropleth mapping, whether straightforward or using the dasymmetric technique, is particularly well served by this method of presentation.

The maps in this edition are mostly choropleth maps, and are based on calculations on data for the districts of the Agricultural Development and Advisory Service of the Ministry of Agriculture, Fisheries and Food (Fig. 12). Each district can be envisaged as occupying a number of squares of the line-printer grid according to its size and shape, and these groups of cells are shaded on the maps with a variety of symbols according to the particular calculation involved and the intervals chosen to divide the results into classes. A line-by-line examination of this kind of map will reveal that the squares along each line 'belong' to different districts, perhaps six to one district, three to the next, five to the next and so on to the end of that line. Moreover, spaces representing sea, or land which is not part of the map, can be left blank or shaded distinctively. Each map is constructed in the same way irrespective of the results of calculation for particular districts and of the intervals chosen to allocate symbols for shading. This makes programming simple, with the single sub-routine required to draw all the maps being modified only by changes in calculations presented and in the intervals chosen for shading. Shading is accomplished by choosing suitable symbols from among those available on the line-printer to indicate increasing or decreasing value, or any other qualitative or quantitative attributer, and printing these symbols in the appropriate grid squares of the line-printer map.

The resulting maps are of reasonable quality and show the intended distributions as well as many traditional choropleth maps drawn by hand or by drum plotter, but it is in their method of construction and production that the real advantages are found. Checks can, of course, be incorporated in most computer calculations to ensure that what was intended was in fact done; similar checks can also be built into the mapping programs to ensure accuracy. Once the system has been set up, and this requires only a few days of programming, any number of maps can be produced with little risk of further complication arising from human errors. Of the maps in this second edition, 247 were produced in 2 hours 45 minutes charged time (or an elapsed time of about 7 hours including printing by the line-printer) on the IBM 360/50 computer at the Edinburgh Regional Computing Centre. Two copies of each map were produced, and, in addition, a superior copy of each map was prepared on special paper in a further 22 minutes of charged time (or an elapsed time of about $3\frac{1}{2}$ hours including printing by the line-printer) on the same machine. This time compares with perhaps 3–4,000 hours for the drawing of similar choropleth maps by hand. Speed is thus a valuable attribute in this type of mapping and one which is continually being improved without pressure from geographers. An IBM 370/155 computer, which has recently been installed at the Computing Centre is resulting in savings of about two-thirds of these times.

There are, however, a number of practical difficulties, though none of them serious, which make for a less efficient production system than that so far envisaged. First, the data for the districts had to be produced in a machine-readable form. Magnetic tapes containing the data were obtained from the Ministry of Agriculture, but these data had to be converted into a form suitable for use on the E.R.C.C. machines. To facilitate the production of maps the individual calculations on the district data were done prior to map production and the results written on to magnetic tape. Two languages were used, FORTRAN to manipulate the data and IMP (a locally developed language) to produce maps. The main reason for this decision was that FORTRAN was the most efficient language for reading the data because of the way in which these were packed, while IMP was the only language which enabled full use to be made of the mapping techniques employed.

Tables and histograms, together with means and standard deviations, were produced for each of the calculations and these facilitated the choice of intervals for shading and choropleth maps. Sometimes the intervals chosen did not prove suitable but it was a simple matter to select one or more alternative intervals and produce further maps from which the most effective could then be selected. The maps were run in batches of about 16, so that over the course of a few weeks, only a few minutes of computer time were required each night. When a final decision had been made on intervals, maps were printed on special output paper, unlined and of good quality, and with a fresh ribbon on the line-printer to get the best possible output for block making.

As for the maps themselves, it will be noted that the grid employed is a square one, and the symbols are not those which normally appear on line-printer output. The printer employed to produce these maps has been specially modified by the E.R.C.C. and IBM for the benefit of the Geography Department of Edinburgh University. Most line-printers operate on a grid of one-tenth by one-sixth of an inch, i.e., the space occupied by each symbol has a width of one-tenth of an inch and the paper advances by one-sixth of an inch for each new line, thus producing a rectangular grid. On many line-printers the space between the lines can easily be changed to one-eighth of an inch, but the grid remains rectangular. A square grid is much more useful for map drawing, and is in fact used in the production of some traditional maps. The closer spacing of lines makes the shading more effective whilst the use of a square grid allows the map to be related to other grids, such as the National Grid which is printed on the maps of the Ordnance Survey. In fact, the grid represented by the symbols on these maps correspond to 5×5 km. grid cells of the National Grid. The print chain on the printer has also been modified and incorporates special symbols chosen to make fullest use of the square grid. The symbols used here consist mainly of horizontal and vertical lines, but overprinting, i.e., the printing of two or more characters in a single location (a facility which is possible on most line-printers) was used for the densest shadings. One disadvantage of these modifications to the printer is that they tend to slow down printing so that they were not always available in busy periods at the Computing Centre. As a result, maps were normally run in the evenings, and long printing runs requiring, say, special paper, were made at weekends.

These problems and their solution, together with inevitable human error in programming, meant that what was in theory a week of concentrated work was spread more thinly over several weeks, with many short runs on the computer rather than one or two long ones. At the end of such a programme of mapping, lessons have been learned, efficiency has been improved and superior quality of production has been achieved, to such an extent that there is temptation to begin again (with perhaps the same result, i.e., discovering even better ways to operate).

Some maps in this edition have been produced in a different manner. The map of leading crops and those indicating crop and other combinations have been made by substituting a distinctive symbol to indicate the actual crop, livestock or enterprise rather than to provide graded shadings to indicate density, although the mapping routine is the same (e.g., Figs. 99–102). Other maps are based on parish data and are rather different (e.g., Figs. 15, 16). They make use of the same grid of 5×5 km. squares of the National Grid, but these are larger than most parishes, at least in the lowlands. A result was therefore calculated for each of the 12,000 or so parishes and an aggregated result produced for each 5×5 km. square, from estimates of those for parishes occurring in whole or in part in each such cell. The resultant map is, of course, much more detailed than those based on districts and is consequently often more difficult to interpret; for each square may well differ from its neighbours, whereas each district map is made up of blocks of uniformly shaded squares. As a result, only a small number of shadings could be used to illustrate such distributions.

Subsequent preparation of maps produced by line-printer for publication involved a reduction in the size of the output, so that lettering produced by the machine was difficult to read. It was decided to produce the final map free of all text and insert that which was required in letterpress at the stage of printing. Reduction in size of the map also limited the number of shadings that could be used. For full-page maps a maximum of six intervals could be effectively used, and for quarter-page maps, only five.

No doubt this method of mapping will be superseded as more sophisticated output devices are developed and storage in the computer ceases to be a serious restraint. For the present, however, the mapping program does what is required cheaply and quickly.

Programming the computer
The flow chart illustrates the main divisions of the map production system. At two main points checks can be made that everything is in order and at the same time

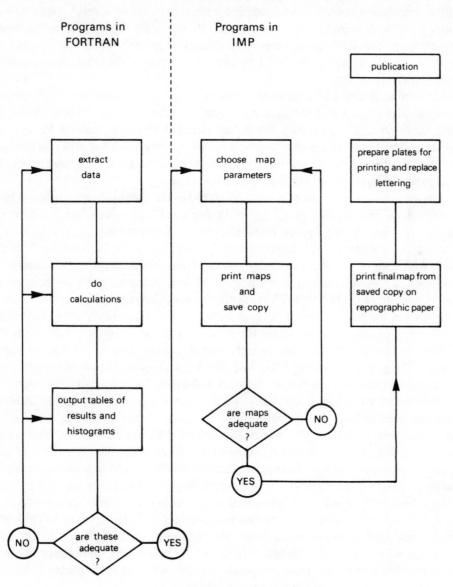

Flow chart for map production

233

refinements can be made of both the question that was asked (the calculation) and the answer that has been given (the map).

The data bank for holding the census data contained a number of arrays, each 380 addresses in length corresponding to the number of districts, and covering all the items used for calculations. Two additional arrays of the same length were used to hold the result of the calculation and the symbols to be used to map the result. A simple percentage calculation routine was used throughout to prepare a result from one or more items as a percentage of other items in the data bank. Where different calculations were required these were programmed separately. The end product in each case was a result accommodated in one of the two additional arrays, with a separate value for each district.

Another simple routine was used to fill the remaining array by examining the results from each district in turn and substituting a symbol chosen for its shading potential, according to intervals for grouping the values detailed in parameters of the routine call. Symbols were either single or composite for overprinting. As the array addressing system was used throughout to correspond to the district numbers, what was produced at this stage was an array of symbols prepared for mapping.

The map used to create the map framework stored in the computer was in fact a line-printer version of the district map shown in Figure 12 in which the hand-drawn boundaries had been replaced by the one-tenth inch grid lines of the line-printer and at a scale at which a one-tenth inch square grid cell represents a 5×5 km. grid cell of the National Grid. This was done simply by placing transparent paper with a one-tenth inch grid over a district map at a suitable scale and redrawing the boundaries to coincide with the lines of that grid. The mapping routine depends on its own data bank which comprises the map framework as detailed on the line-printer grid. This data bank is an array of numbers indicating the pattern of the map by grid cells from the top left-hand corner of the map, left to right and down the map to the bottom right-hand corner. The grid cells belong either to a district which can be numbered or to the sea or unsurveyed land, and these were detailed in the data bank by pairs of integers to indicate their occurrence. There are many ways of doing this and a series of CAMAP programs has been written to cope with different kinds of framework. In the district maps produced in this volume, the use of the integer $\emptyset$ to indicate spaces and -1 to indicate a new line, in addition to the district numbers 1 to 380, is sufficient. All integers are in pairs, except -1. These pairs consist of either a $\emptyset$ or a district number followed by another integer indicating the number of grid cells at that point in the line of the map. The number -1, indicating a new line, follows the last occurrence of a district along the previous line. A unique number, such as -2, can be used to close the map framework and return control from the mapping routine. The district map framework fell within the 130 character width (13 inches) of the line-printer, so that extra programming was not necessary to produce maps in two or more parts.

Part of the data bank of a map framework might read:

... 16 10 57 4 16 1 59 3 $\emptyset$ 4 59 2 -1 $\emptyset$ 32 ...

These figures indicate that at this point in the construction of the map there were 10 symbols representing district 16, followed by 4 representing district 57, 1 representing district 16, 3 representing district 59 followed by 4 spaces and then another 2 symbols representing district 59. This is followed by a new line, then 32 spaces and so on until the map framework is completed.

To operate, the map routine needs an associated array of symbols set up for mapping and these are printed at the appropriate points in the map as already described. The most convenient way to print these symbols is a line at a time, and to print each line repeatedly until all composite symbols for overprinting have been dealt with. This can be done in several ways depending on the complexity of the symbols used and the extent of overprinting required. In the district maps produced here, the map routine constructs the map line by line, filling an array of 130 addresses representing the 130 character positions of the line-printer line and made up of either spaces or shading symbols appropriate to the particular line of the map. Composite symbols are packed into the individual addresses of the line in such a way that they can be unpacked one symbol at a time. These are printed individually at each printing of the line until the unpacking and overprinting is completed. The map routine then moves on to the next line, indicated by a −1 in the framework data, and makes use of the same 130 length array for printing that line, and so on until the map is completed.

As noted already, this method of printing a map obviates the need for two co-ordinates detailing the location of each grid cell or data point in the map, and saves much more space in the computer by repeatedly using the same small line array, so avoiding the use of the two dimensional graphics array common to most computer mapping programs.

The more sophisticated techniques used in mapping from the 12,000 or so parishes, and the reduction of the result to the 5×5 km. grid as illustrated by eight such maps included in the atlas, are not here described in detail although they were developed by the same basic approach to computer mapping.

Appendix III

SELECTED STATISTICS

TABLE 1

Percentage of crops and grass under tillage, crop combinations and proportions of tillage under six leading crops, in rank order; livestock units per 100 acres of agricultural land, livestock combinations, and proportions of livestock units accounted for by five leading livestock (in rank order); man-days per 100 acres of agricultural land, enterprise combinations and proportions of man-days under different enterprises (in rank order). The figures given here may not always agree with those on the maps, partly because of rounding, but chiefly because adjacent districts have sometimes been amalgamated for mapping purposes. The following abbreviations are used. *Crops:* b, barley; c, cabbage; d, rye; e, beans for stockfeed; f, French beans; g, green peas; h, hops; i, carrots; j, broad beans; k, kale; m, mixed corn; o, oats; p, potatoes; q, winter cauliflower; r, rape; s, sugar-beet; t, turnips, swedes and fodder-beet for stock feeding; u, savoys; v, other crops for stockfeeding; w, wheat; x, summer cauliflower; y, peas for harvesting dry; z, Brussels sprouts. *Livestock:* b, beef cattle; c, crops; d, dairy cattle; f, poultry; p, pigs; s, sheep. *Enterprises:* b, beef; c, crops; d, dairying; f, poultry; h, horticulture; p, pigs; s, sheep. Letters in capitals are those included in combinations; other crops, livestock or enterprises are shown in lower case.

		Crops								Livestock							Man-days										Area	
District	Percentage under tillage	Crop combinations	Proportion of tillage under six leading crops in rank order						Livestock units per 100 acres	Livestock combinations	Proportion of livestock units under five leading livestock in rank order					Man-days per 100 acres	Enterprise combinations	Proportion of man-days under seven enterprises in rank order							Crops and grass 000 acres	Rough grazing 000 acres
1	49	BWotpr	57	18	12	7	2	2	23	BSdpf	53	33	6	4	3	274	CBSdpfh	43	27	20	5	3	2	1		120	49
2	36	BWotpr	57	16	12	9	2	2	27	BSdfp	53	34	6	4	4	277	CBSdfph	32	31	23	6	3	3	2		108	41
3	11	BTOrwk	50	16	16	7	1	1	15	SBdfp	50	44	3	2	0	113	SBcdfph	47	38	9	4	2	0	0		60	194
4	5	BOtrwk	46	25	13	6	4	3	21	BSdfp	45	36	17	2	0	113	BSDcfph	37	34	22	4	2	0	0		46	86
5	22	BOWtrp	63	12	10	5	2	1	33	BDSfp	47	28	20	3	2	309	BDSCfhp	32	30	16	16	3	2	1		63	23
6	23	Bwotrp	64	12	12	6	2	1	34	BSdfp	58	23	12	4	3	308	BCSDhfp	38	19	19	13	5	4	2		99	18
7	50	Bwoptc	69	13	6	4	3	1	28	BDFSp	42	20	16	14	8	403	CBDHfsp	38	19	15	9	8	4	0		86	3
12	12	BOtmpk	55	14	11	8	3	2	33	DBsfp	40	38	18	3	1	278	DBScfph	44	28	16	9	2	1	0		66	52
13	18	BOMtpw	54	14	11	9	4	2	57	DBsfp	57	31	5	4	3	575	DBcfshp	56	19	12	4	4	3	2		74	5
14	14	BMOtpw	47	19	13	9	6	3	57	DBsfp	55	30	9	5	1	516	DBcsfhp	56	20	9	7	5	2	1		75	13
15A	22	BTompw	57	14	10	7	5	4	33	DBSfp	43	29	19	6	3	268	DBCSfph	43	19	16	15	5	2	0		73	61
15B	11	BTompw	60	12	10	7	3	3	42	DBsfp	50	27	18	4	2	227	DBScthp	53	18	15	7	3	3	1		62	35
16	9	BMtopw	56	12	10	9	4	2	31	DBSfp	45	29	23	2	1	165	DBScthp	50	21	19	6	2	1	1		70	71
17	10	BOMTpk	36	21	12	10	9	5	29	DBSfp	45	26	24	2	2	143	DSBcfph	49	21	18	8	2	1	1		67	87
18	14	BMTowp	57	12	12	8	4	3	29	DSBfp	43	27	26	3	1	124	DSBcfhp	45	23	18	9	3	1	1		69	90
18A	2	BOpkrm	51	20	7	6	4	4	29	SDBfp	37	36	21	3	3	151	DSbfphc	43	33	17	2	2	2	1		32	58
19	3	BOKptr	41	23	10	6	5	3	47	DSBfp	52	21	18	8	1	308	DSbfhcp	58	18	13	6	2	2	1		53	28
19A	3	BTMokr	47	15	15	11	5	3	29	DSBfp	38	35	26	1	0	115	DSBcpfh	45	31	20	2	1	1	0		51	65
20	44	Bwoptm	63	14	10	6	3	1	32	BDSfp	38	30	14	12	7	416	CDBsfph	36	24	19	9	7	4	2		56	3
21	31	Bwoptx	61	10	10	7	4	2	29	BDSfp	41	27	14	14	4	255	CDBFshp	26	23	23	11	10	5	3		56	15
22	60	Bwoptc	64	14	9	8	3	1	32	DBPFs	41	22	19	16	2	547	CDhbpfs	41	26	9	8	7	1			25	1

No.	n	Code	a1	a2	a3	a4	a5	a6	a7	Code	b1	b2	b3	b4	b5	N	Code	c1	c2	c3	c4	c5	c6	c7	c8	c9
23	55	BWoptm	58	16	12	7	2	1	34	BDFps	30	29	20	12	9	579	CFDbpsh	33	25	18	11	5	4	3	61	2
24	48	BWoptm	63	17	9	6	1	1	36	BDPsf	37	30	15	10	8	502	CDBHpsf	31	22	17	11	8	6	5	52	1
25	20	Bowptr	63	15	10	4	2	2	48	DBspf	39	33	14	9	5	438	DBCspfh	41	22	12	10	6	5	3	46	5
25A	4	Btporm	61	10	6	6	5	5	20	SBDfp	42	34	20	3	12	85	SBDcfph	39	27	25	3	3	1	1	38	82
26	72	Bwtops	67	16	6	4	3	2	25	BPSDf	28	22	20	18	6	432	CBDpsfh	53	12	11	9	8	6	5	103	2
27	68	BWtogp	58	20	5	4	3	2	22	BSDpf	34	24	21	11	10	411	CBDsphf	53	11	8	6	6	5	4	108	3
28	72	BWGpso	48	13	10	7	6	5	26	PBDsf	37	24	17	17	5	553	CPdbhfs	56	12	12	8	8	6	6	82	1
29	68	BWGspo	42	19	9	8	7	7	26	PBDFs	26	26	26	10	10	584	CHDpbfs	51	13	11	8	4	2	3	77	3
30	72	BGWzpo	42	19	9	8	7	7	36	PBDfs	51	19	14	14	6	1046	HCpdbfs	53	22	10	5	7	3	1	62	3
31	68	BWopgt	40	18	17	5	4	4	32	PDBsf	36	25	23	10	6	487	CDPBfsh	40	17	10	10	5	5	5	77	1
32	73	BWgoyp	53	28	7	2	1	2	40	PBdfs	56	17	13	8	6	572	CPDhbfs	38	27	10	10	7	6	6	102	0
33	72	BWGoty	44	21	20	4	3	2	26	BPDSf	35	25	17	16	7	466	CBPDfhs	47	14	10	10	7	6	0	22	131
34	9	Bwpkor	68	9	7	4	4	4	18	SBDfp	40	30	25	3	2	60	SDBcfph	35	31	23	6	3	2	2	49	10
34A	44	Bwopmt	69	11	7	7	4	3	41	DPBFs	30	29	16	12	12	465	CDBFpsh	29	26	16	10	9	8	2	79	79
35	9	Bwokpt	70	11	7	6	4	3	25	DSbpf	43	34	19	13	8	136	DSbcfph	49	28	14	5	2	8	2	52	15
35A	39	Bwpotk	66	10	6	7	3	1	39	BDPsf	34	31	14	13	13	349	CDBpsfh	28	28	17	9	8	8	2	69	2
36	62	BWpsot	60	12	11	7	4	2	35	BPFDs	32	20	20	17	8	647	CFBdpsh	46	20	11	9	8	4	2	90	2
36A	58	BWsopm	59	11	8	7	7	3	34	BDPfs	36	31	16	9	10	529	CDBpfsh	46	20	14	12	5	4	3	70	35
37	47	BWpotc	63	16	7	7	2	1	30	DBSpf	33	27	20	10	9	262	PCDBshp	26	24	23	13	5	5	5	63	45
38	24	Bowtpr	63	12	12	3	2	1	25	BSDpf	44	29	16	7	10	113	DBCSpfh	45	19	13	13	4	0	5	42	38
39	46	Bwtomp	63	13	7	5	3	3	24	DBSPf	37	25	22	12	4	164	CBDSpfh	34	20	19	16	7	8	5	65	14
40	49	Bwotpm	65	12	6	6	2	2	33	BDPsf	33	28	16	16	8	356	CDBHspf	29	25	13	11	9	4	3	66	3
41	61	BWspot	59	14	7	7	5	2	29	DBSpf	37	28	20	10	5	502	CDBpsfh	50	17	14	8	4	3	2	71	8
42	43	BWopmt	63	18	10	5	1	1	37	DSBfp	44	25	13	12	3	381	DCBspfh	37	27	14	8	7	5	0	60	0
43	0	TPKOrm	48	14	12	11	9	2	26	DSbfp	39	36	21	3	1	122	DSbfphc	45	32	16	5	1	2	0	62	119
44	0	TRKMZc	32	16	16	10	8	8	41	DSbpf	58	22	13	6	5	330	DSbfphc	65	19	10	2	2	0	0	64	36
45	0	KTrvp	52	22	14	8	4	0	27	DSbpf	46	33	18	5	2	140	DSBfpch	55	29	13	6	2	0	0	48	67
46	4	BKowrv	59	14	7	4	3	3	31	DBSfp	49	21	17	7	7	182	DSBfpch	56	16	15	8	5	2	0	42	48
47	49	BWopst	61	13	8	8	4	2	40	DBPfs	34	32	16	10	6	560	CDBfphs	34	23	15	7	8	7	4	55	3
48	5	Btokrw	72	4	4	3	3	4	44	DSbfp	53	17	17	8	8	284	DSbfphc	57	14	11	7	5	4	2	44	21
49	52	BWopst	54	19	9	7	5	2	32	DBPsf	35	34	16	8	7	526	CDBHpfs	40	21	13	12	4	4	4	101	2
50	71	BWpocs	50	18	7	7	3	3	22	BPDfs	29	29	23	15	4	695	CHdpfbs	41	31	8	7	6	1	1	56	1
51	0	KCxtps	54	25	12	4	3	2	47	DBFsp	53	16	12	10	9	337	DFbsphc	58	11	10	8	7	6	0	26	13
52	1	CBXKut	29	21	17	14	4	4	58	FDBps	34	28	17	13	8	322	DFBpshc	34	33	12	10	2	0	0	27	18
53	23	BQCXop	52	9	4	8	4	4	65	DPFBs	36	25	21	16	2	821	DHFPbcs	32	25	14	13	8	7	1	34	2
56	85	BWGpso	34	23	18	8	6	5	16	PBFDs	39	30	18	9	4	660	CHpbfds	56	25	7	5	5	2	1	90	1
59	73	BWGPpf	31	20	20	9	4	3	20	BDPfs	35	29	22	10	3	543	CHdbpfs	57	15	11	8	5	3	1	98	1
60	61	BWpocq	57	14	8	6	3	3	37	DPFbs	33	25	21	20	2	605	CDHPfbs	33	21	17	10	8	7	1	36	1
61	6	Bmpotk	60	8	8	6	6	3	37	DBFps	42	20	19	11	9	251	DFBpshc	47	16	14	8	7	5	3	29	25
62	21	Bowpmt	62	10	9	9	6	3	48	DBFes	44	25	15	9	6	487	DBCFpsh	45	16	13	13	6	5	3	38	7
65	67	BWgpoj	56	18	8	7	3	1	30	DBPFs	31	25	23	20	2	535	CDHFbps	42	19	10	10	9	9	1	64	3
67	5	BOmpkt	44	23	9	8	4	4	38	DBSfp	49	24	20	3	3	205	DSBcfph	56	18	16	4	3	2	1	59	47
68	2	BPOckt	54	10	9	5	5	5	45	DBSfp	51	19	18	9	3	266	DSbfhpc	57	16	13	7	3	2	1	71	41
69	11	BWOpmx	50	13	10	8	8	3	106	DFPbs	43	28	15	10	4	1058	DFHpbcs	44	20	14	10	6	3	3	168	14

District	Percentage under tillage	Crop combinations	Proportion of tillage under six leading crops in rank order						Livestock units per 100 acres	Livestock combinations	Proportion of livestock units under five leading livestock in rank order					Man-days per 100 acres	Enterprise combinations	Proportion of man-days under seven enterprises in rank order							Crops and grass 000 acres	Rough grazing 000 acres
70	3	BCGOpx	45	15	8	8	7	5	52	DFPbs	46	21	13	12	8	406	DFpbshc	51	18	9	8	7	5	1	113	54
71	63	BPZWIC	40	10	8	8	7	7	37	FDPbs	31	26	24	17	2	943	HCDfpbs	43	25	11	10	6	4	1	169	14
76	24	BWOpmc	48	12	11	7	7	4	65	DFpbs	58	15	13	12	1	839	DHfcpbs	50	18	10	9	6	6	1	56	2
76A	19	BMWOpc	46	14	13	12	5	2	65	Dbfps	65	17	10	8	1	741	Dbcfphs	65	10	7	6	5	4	1	66	1
77	16	BMWopk	55	16	14	4	4	3	71	Dbpfs	69	12	10	8	1	768	Dbcfphs	71	7	6	6	5	4	1	65	1
78	39	BPWomk	47	19	12	10	7	1	53	DBfps	62	15	13	9	1	775	DCfbhps	47	27	9	7	5	4	1	54	1
78A	38	BPMwoz	52	12	11	10	7	2	52	DFBps	54	18	15	11	2	850	DHCfbps	36	26	18	8	6	5	1	52	2
79	18	BMPowk	43	22	16	6	6	2	62	Dbfps	69	11	9	6	4	727	Dcbfhps	64	11	6	6	6	4	3	48	5
79A	4	BMPkot	46	21	11	6	4	4	43	DBFsp	51	18	13	11	7	422	DBFshpc	55	12	12	9	6	5	3	34	17
79B	20	BMPwok	42	22	11	9	9	1	65.	Dbpfs	72	13	8	7	2	791	Dcbfphs	67	11	7	6	5	4	1	53	1
80	0	Vpct	77	17	4	2	0	0	46	Dbsfp	67	22	7	2	2	440	Dbsfpch	76	15	6	2	1	0	0	33	6
80A	2	BMPkoc	33	22	19	8	7	4	62	Dbfps	68	16	7	6	3	632	Dbfphsc	73	10	5	4	4	2	1	34	3
81	3	BOVkwp	51	19	13	8	4	2	55	Dbsfp	70	20	4	3	3	549	Dbsfpch	77	13	3	2	2	1	1	32	2
81A	24	BMWpok	51	17	15	7	5	2	56	DBpfs	62	18	11	8	2	633	DCbpfhs	60	14	10	7	5	3	1	45	1
82	22	BWmops	48	24	8	7	6	2	59	DBfps	58	23	9	7	2	663	DBchfps	57	13	11	7	6	4	2	53	0
82A	16	BWomkp	52	26	8	6	5	3	60	Dbfps	66	19	6	6	3	633	Dbcfphs	67	12	7	5	4	3	3	51	1
83	30	BWmopk	55	23	6	6	5	2	52	DBfps	59	22	8	7	5	563	DCbfpsh	57	17	12	5	4	3	1	55	1
83A	37	BWopms	44	28	8	7	5	4	44	DBpsf	59	26	7	5	4	556	DCbpfsh	49	26	13	4	3	3	2	62	1
84	54	BWPouc	39	24	11	5	4	4	34	DBfps	52	23	11	9	5	649	CDHbfps	34	28	19	8	5	4	2	67	2
85	54	BWPsoc	49	16	13	9	5	2	82	DBpfs	28	25	21	21	6	676	CHDbfps	38	24	15	8	7	6	3	59	2
86	22	BMWokp	53	16	12	7	6	3	61	DBpsf	58	21	9	8	6	627	DBcpsfh	58	13	11	6	5	4	3	60	2
87	31	BWmspo	54	12	9	9	5	4	63	DBfps	58	15	13	12	2	722	DCbfphs	55	18	9	8	7	2	1	67	1
88	38	BWsomp	50	19	9	7	6	5	51	Dbpsf	66	14	13	4	3	646	DCbpfsh	53	23	8	7	4	3	2	86	1
89	56	BWSpom	43	19	16	13	3	2	40	DBPfs	37	24	21	11	7	693	CDbpfhs	46	22	9	8	6	6	3	71	1
89A	59	BWSpoe	41	21	20	9	4	1	34	DBPsf	49	24	17	6	4	726	CDHbpsf	45	24	15	8	5	2	1	34	0
90	23	BOWtrm	47	18	17	6	4	3	40	BSdfp	47	35	11	5	2	376	BSCdfph	31	27	18	11	11	2	0	80	8
90A	23	BWomst	45	26	10	9	2	2	49	DBSpf	43	26	15	8	8	412	DBCSpfh	46	18	13	12	6	5	0	53	8
91	36	BWosmp	47	24	9	6	5	3	41	DBspf	51	24	12	9	5	394	DCBspfh	44	24	14	8	5	4	1	64	9
91A	47	BWsopm	50	17	9	9	6	2	41	DBPSf	31	25	18	14	12	557	CDBPsfh	35	24	12	10	8	6	5	64	1
91B	36	BWotmk	47	27	14	4	2	1	38	BSDpf	32	27	24	9	4	415	CDBSfph	25	25	18	17	7	5	1	44	2
91C	24	BWOtkr	44	26	19	8	2	1	39	BSDpf	42	26	25	4	3	369	DBSCfhp	25	24	19	16	9	4	3	58	5
92	31	BWOtpm	35	27	18	6	6	3	40	BSdfp	47	29	15	6	3	479	CBSHDfp	26	23	17	15	11	5	2	74	2
93	31	BWOtpr	41	24	18	8	4	2	41	BSDfp	42	24	19	11	4	496	CBDSFhp	24	21	16	15	13	9	2	65	2
94	19	BWOTtp	35	23	15	10	7	2	40	BSDpf	39	36	17	4	4	353	SBCDfhp	26	24	20	16	9	3	3	59	7
95	30	BWOhtp	34	29	19	5	4	3	40	BSDpf	37	27	23	8	5	608	CDBHSfp	33	18	16	14	11	4	4	61	1
96	35	WBOpht	35	33	11	7	4	3	41	DBspf	37	35	13	8	7	751	CHDBsfp	29	25	22	12	5	3	3	63	1

No.	n	BW pattern	BW data	DB n	DB pattern	DB data	CH n	CH pattern	CH data	final
97	36	BWspot	48 19 8 7 6 4	42	DBSpf	31 29 22 9 9	641	CHDBSfp	24 21 21 50 8 11 21	1
98	42	BWOhps	35 29 10 8 6 4	34	BDSpf	40 26 16 11 7	814	CHDbspf	50 16 11 62 4 11 16	0
101	29	BWOemp	44 37 9 8 1 1	42	DBSPf	33 31 13 13 10	493	DCBHSpf	28 17 17 23 3 5 17	1
102	49	ZBWcox	26 26 23 6 4 3	31	DBFPs	26 25 19 17 13	1099	Hcdfbps	67 10 7 81 3 5 17	1
103	30	BWOpsc	47 18 9 5 4 4	51	DBFPs	36 23 19 15 7	652	DHCFBps	30 19 15 50 6 12 5	1
104	33	BWOsph	41 27 11 6 4 3	37	BDSpf	33 25 24 13 5	681	HCDBspf	31 26 14 57 5 10 14	2
105	34	WBohcs	33 33 8 6 3 4	37	DBFps	39 27 12 12 11	764	HCDbfsp	31 27 19 61 3 7 11	1
106	47	BWSZcp	33 17 14 8 8 6	42	DFBPs	25 24 23 18 10	827	HCDFbps	34 26 13 55 4 6 13	1
109	53	BWpomx	47 26 9 4 2 2	38	DBFPs	50 21 14 13 2	577	DCbfhps	35 34 9 52 6 9 9	0
110	34	BWOmpk	52 23 13 7 2 2	49	DBFPs	38 23 22 13 4	570	DCFBHps	35 16 14 43 1 10 14	1
111	41	BWopms	49 29 9 7 2 1	36	DBspf	43 32 10 9 6	477	DCBshpf	35 30 15 59 3 6 8	1
112	52	BWopec	44 33 11 6 2 1	30	DBpsf	38 31 11 11 10	457	CDBhsfp	37 26 13 72 5 6 6	0
113	47	WBzoep	37 37 8 7 1 2	31	DBSfp	31 31 16 10 10	457	CDHBsfp	28 21 15 74 5 6 6	1
114	40	BWoemy	43 41 10 3 1 1	31	BDSfp	35 30 18 9 8	345	CDBSfph	29 27 19 67 5 5 5	0
115	44	BWozce	41 31 10 5 2 2	31	DBFsp	44 26 12 10 7	542	HDCbfsp	30 26 20 62 3 5 5	17
116	3	BKROmp	41 18 16 15 4 3	40	DBsfp	53 31 12 3 1	355	DBsfcph	62 22 11 61 1 2 2	72
116A	1	BRkotp	40 33 11 8 4 2	24	DBSpf	37 26 25 7 5	114	DSBpfch	47 23 20 39 0 4 11	9
117	22	Bomwpk	65 9 9 6 4 3	39	DBpfs	50 25 12 11 2	480	DHBCfps	44 15 14 35 1 8 8	1
117A	51	BWopmc	57 15 9 9 3 1	39	DBfps	43 26 15 14 3	582	CDBhfps	32 31 12 52 6 3 9	1
118	18	BMowkp	58 11 10 9 4 3	54	DBfps	62 23 9 7 3	583	DBcfpsh	62 14 9 44 2 2 10	1
118A	42	BWzomc	51 15 8 5 4 4	41	DBfsp	54 23 14 4 4	641	DHCbfps	37 22 19 46 3 3 7	1
119	34	BWompk	51 23 9 6 4 3	50	Dbfsp	62 22 7 4 4	611	DCbhfsp	54 17 11 58 2 2 6	2
120	10	BOmwkp	56 12 11 11 5 2	55	BDPFs	69 19 5 5 2	562	Dbfcsph	71 12 6 56 1 0 4	1
121	68	BWgspi	40 22 12 7 6 4	26	BDPFs	32 24 21 18 5	581	CHDbfps	48 16 12 90 0 7 9	0
122	68	BWsope	49 29 8 6 3 2	27	FDbps	34 29 20 11 3	475	CDBpfhs	50 17 13 86 2 7 7	1
123	68	BWsgpo	58 13 8 8 5 4	59	DBPfs	62 14 11 10 3	845	FCdhbps	39 31 11 87 1 6 6	0
124	56	BWsopy	49 30 9 4 3 1	32	DFBPs	35 34 16 9 6	533	CDHBpfs	37 22 15 77 3 3 5	1
124A	59	BWspoz	51 21 6 5 5 3	37	DFBPs	32 27 21 17 3	657	CHDFbps	32 22 20 41 1 8 13	1
125	80	BWGspo	35 25 15 11 6 3	15	BPFds	33 29 22 12 4	518	Cdbfphs	71 6 6 109 2 5 11	1
126	85	BWPSyz	31 30 11 10 5 3	16	BDPsf	41 26 14 12 7	624	Chfbpds	69 7 5 64 3 4 6	1
127	71	BWgspo	44 28 7 8 4 3	17	BPDFs	38 19 16 15 12	393	Cdbspfh	62 11 4 123 3 3 7	1
128	78	BWPSoe	33 31 8 8 5 4	14	PFBds	37 30 22 10 1	531	Chbdfps	65 14 5 99 3 4 11	2
129	90	BWGPsi	24 22 20 9 7 6	11	FBPDs	32 29 15 15 9	678	CHpfbds	63 24 4 68 0 3 5	1
130	78	BGWspz	29 25 23 7 6 3	19	BFPps	41 22 16 13 9	654	CFhbdps	51 22 11 103 2 5 4	1
131	76	GBWpso	33 23 4 6 4 2	17	BPFps	36 22 17 13 12	465	CHbdfps	56 14 10 100 3 5 10	1
132	75	BWGPso	30 23 12 6 4 4	19	FBPds	52 18 16 11 3	573	CHbdfps	56 14 10 80 4 4 9	1
133	75	BWGpso	38 24 12 7 7 3	35	BDSfp	48 21 13 10 8	666	CFbdphs	48 27 6 101 5 3 6	1
134	68	BWGypo	31 26 23 5 4 3	17	BDpfs	43 23 17 13 5	379	CBdhsfp	59 14 10 78 1 4 6	2
135	71	BWGCPX	19 14 14 9 9 4	19	BDFPs	31 20 18 16 15	718	CHbdfps	47 32 7 72 3 3 3	1
136	70	BWsgpo	46 23 7 5 4 2	21	BFDPs	29 25 21 13 13	487	CFdbsph	56 11 9 87 4 6 9	1
137	75	BWGspo	40 23 15 6 5 3	20	BFDPs	32 23 22 20 4	488	Cfdbhsp	60 11 8 101 4 5 8	1
137A	79	GBWzsp	37 27 23 3 3 3	17	BSdpf	49 24 14 9 5	467	CHdfbps	55 14 11 22 5 4 11	0
138	59	BWokse	55 26 5 3 3 3	20	BDpfs	39 36 10 8 7	326	CBsdpfh	54 18 7 82 3 3 9	0
139	45	BWoype	42 32 8 5 3 3	34			469	CDBHfps	29 27 17 64 4 7 11	1

District	Percentage under tillage	Crop combinations	Proportion of tillage under six leading crops in rank order	Livestock units per 100 acres	Livestock combinations	Proportion of livestock units under five leading livestock in rank order	Man-days per 100 acres	Enterprise combinations	Proportion of man-days under seven enterprises in rank order	Crops and grass 000 acres	Rough grazing 000 acres
139A	46	BWopzm	53 24 9 5 2 1	38	DBfps	47 28 11 10 3	516	DCBhfps	36 29 13 8 7 5 2	58	1
140	46	BWogep	45 33 7 4 3 1	38	DBfsp	40 31 12 10 7	410	DCBfsph	34 32 16 7 6 4 1	84	1
141	37	BWoeyk	43 36 10 5 2 1	36	BDSpf	48 25 19 5 3	371	CBDShpf	26 26 24 14 4 3 3	84	0
142	32	BWoepk	50 28 13 2 2 2	45	BDpsf	37 37 11 10 6	489	DBChspf	36 20 18 8 7 4 3	59	0
143	48	BWopem	50 27 11 3 2 2	37	DBpfs	41 22 14 13 10	492	DCBpfhs	39 28 14 7 5 5 3	71	1
144	63	BWoype	47 30 7 4 3 2	22	BDPSf	43 23 16 13 6	378	CBDspfh	52 14 13 6 6 4 3	107	1
145	58	BWoepy	47 32 6 3 3 2	26	BDSpf	46 21 21 8 5	375	CBDsphf	45 17 16 9 6 4 3	121	1
146	43	BWoepk	46 36 8 2 2 1	32	BSDfp	40 27 13 11 10	347	CBDSfph	33 22 19 15 6 3 2	94	0
147	52	BWopky	56 25 6 3 2 2	31	DBSpf	34 31 19 10 7	406	CDBshpf	35 21 17 8 8 6 5	46	1
148	48	BWoeky	46 37 10 2 2 1	30	DBSpf	37 29 27 4 2	362	CDBSpfh	34 28 16 13 6 4 1	109	0
153	93	WXGYZC	16 15 12 10 8 8	9	PBFsd	36 34 27 2 1	1486	HCpbfds	65 30 2 1 0 0 0	59	1
154	91	XCGWZP	20 14 13 11 8 8	11	BPFds	46 34 9 8 3	1574	HCbpfds	67 27 2 2 1 0 0	41	2
155	93	WYGBSP	22 21 11 10 9 9	7	PBdfs	55 21 14 8 2	1873	Hcpbdfs	73 24 1 1 0 0 0	43	1
156	95	YWGPBs	25 20 18 10 9 7	7	PBfsd	49 23 22 3 2	1525	HCpbfsd	63 34 4 2 1 0 0	55	1
157	94	WYBGPs	25 23 16 13 9 7	10	PFBsd	38 28 22 8 5	824	CHpfbsd	58 34 4 2 2 0 0	40	0
158	84	BWISyg	44 14 13 12 4 3	11	PBDfs	43 21 17 10 9	589	CHpdbfs	54 31 5 4 4 2 1	123	3
158	78	BIWSyd	42 16 12 11 5 3	18	PBDfs	40 20 19 18 3	527	CHpfdbs	45 23 10 9 6 4 3	51	11
160	85	BWISPy	22 21 17 14 11 7	14	PBFDs	57 21 20 2 0	922	CHfpdbs	50 36 5 4 3 2 0	102	2
161	89	YWGBSf	20 20 17 14 10 7	8	PFBds	34 28 19 18 2	1381	HCfpbds	59 34 3 2 1 0 0	51	0
162	86	BWSGpi	38 15 12 12 5 4	20	PDFBs	36 32 18 12 2	709	CHdpfbs	53 22 8 7 6 4 0	80	2
163	83	BGWSfp	30 19 14 13 8 6	31	FPDbs	45 28 14 13 1	907	CHFpdbs	44 23 16 7 6 3 0	89	4
164	80	BWSgif	45 15 13 6 6 3	26	PDbs	32 27 26 13 2	688	CFHPdbs	45 15 14 11 11 4 0	81	3
165	67	BGWSfp	30 28 14 12 4 3	31	DPBfs	40 30 21 8 2	731	CHDfpbs	41 20 13 10 8 7 1	102	4
166	80	BSWgyo	44 18 12 5 4 3	23	PDBfs	34 34 20 11 0	594	CHDpbfs	51 16 12 11 6 4 1	104	3
167	84	BWsgoe	51 19 9 9 4 2	35	DPFbs	29 26 24 19 2	648	CDFPhbs	39 19 14 13 10 4 0	84	2
168	81	BWScio	46 16 13 3 3 3	18	PBDFs	69 14 12 4 1	670	CHFdpbs	48 18 15 7 6 5 1	61	3
169	77	BISwif	48 14 13 10 3 3	59	Fpdbs	40 28 18 11 3	1131	FCHdpbs	47 24 15 6 5 2 1	30	1
170	83	WBGYSp	25 20 17 11 10 8	17	PBDfs	43 27 16 13 0	802	CHpdbfs	50 33 6 4 4 2 0	19	1
173	73	GBWsof	33 31 15 6 2 2	30	DPBfs	45 25 17 12 1	630	CDHpfbs	40 21 17 9 8 5 0	78	2
174	78	BWgyso	48 18 7 7 7 3	33	PDFbs	59 17 17 7 0	585	CPDFhbs	42 18 15 11 10 5 1	71	2
175	86	BWgsyf	43 19 10 9 4 4	44	PFdbs	48 19 19 14 1	718	CPFdhbs	41 23 17 10 6 3 0	77	1
176	81	BWsheg	48 21 9 7 5 4	28	PFDbs	42 24 18 14 2	634	CHPdfbs	44 20 14 8 8 5 1	65	1
177	80	BWSgyi	41 19 9 7 5 4	24	PDBfs	44 25 17 8 7	637	CHpdbfs	48 21 11 9 5 5 1	94	2
178	75	BISWpd	40 19 14 13 4 3	17	PBDsf	52 16 16 15 3	612	CHpbdfs	46 29 8 5 5 3 2	80	6
179	86	BWsyeo	43 29 9 6 5 3	23	PDBfs	51 19 16 11 3	531	CPhdfbs	57 15 10 7 5 5 0	76	1
180	86	BWesoy	44 31 9 6 4 2	14	PBFds	51 19 16 11 3	461	CHpfbds	58 18 11 5 4 1	104	1

Entry		Code							Code								Code								
181	85	BWSyeo	46	25	13	5	3	2	PFDbs	32	50	23	15	11	2	660	CPHFdbs	48	16	12	12	7	4	77	1
184	83	BWSipy	36	27	11	4	5	2	BPFds	9	31	30	26	10	4	587	CHfbpds	59	28	4	4	3	2	61	1
185	82	BWezyp	38	34	8	4	4	3	PBDfs	16	46	25	17	9	4	592	CHpdbfs	40	39	9	5	4	2	72	1
186	88	BWsczg	44	26	8	6	4	2	PFBDs	16	44	22	15	15	4	522	CHpfdbs	56	20	9	5	5	3	98	1
187	93	WSBPYi	30	19	15	15	9	8	PBDfs	8	45	26	17	12	0	956	CHpbdfs	59	34	3	3	1	1	58	0
188	79	BSWpyi	45	18	18	7	4	3	PBDfs	19	35	32	25	6	6	744	CHpbdfs	46	34	4	6	1	0	9	1
189	89	WYBPGS	23	21	14	12	9	2	PBfsd	12	62	27	9	1	0	1410	HCpbfsd	58	35	1	1	1	0	46	1
190	94	WPSBGy	30	16	16	14	11	9	PFbds	23	47	45	6	1	0	981	CHfpbds	65	18	8	7	1	0	58	1
191	85	WBIPSy	31	17	13	12	12	5	PFBds	18	37	31	20	12	0	910	CHpfbds	51	34	5	5	3	3	46	1
192	82	BWSPgy	32	28	11	8	7	5	BPFDs	16	28	25	21	20	5	567	Chfdpbs	69	7	7	6	5	0	96	2
193	81	BWyezp	36	30	8	6	6	4	PBDfs	13	35	28	16	13	7	444	CHpfbds	59	15	7	6	6	5	58	0
194	84	WBZyes	32	30	14	12	5	5	BPDfs	16	41	26	26	5	1	449	CHdbpfs	63	20	6	6	4	6	62	1
195	85	WBSPfe	32	24	19	5	4	3	PFbds	22	43	29	14	13	3	866	CHfpdbs	54	22	14	6	3	2	55	0
196	77	BWZpgo	38	26	15	9	5	3	PFBDs	17	27	26	24	18	4	657	HCpfdbs	37	37	7	3	6	1	100	1
197	83	ZBWcps	36	21	6	5	3	2	BPDfs	33	31	30	21	15	2	1197	Hcbpdfs	67	21	3	3	6	0	62	0
198	65	BWzoep	59	22	5	4	1	1	PBFDs	20	26	25	25	21	2	534	CHDPBfs	31	21	14	13	11	6	52	2
199	78	BWzeop	46	31	5	4	1	1	PDBFs	26	28	24	23	23	2	503	CHFdpbs	44	19	10	10	10	0	85	2
199A	68	BWeopk	64	23	5	5	1	1	DPBFs	31	35	23	20	20	2	2	CDFHbps	39	22	11	5	9	2	30	3
200	59	BWeops	50	31	7	3	1	1	DPBfs	37	39	25	20	14	2	1020	HCdpbfs	57	16	13	4	5	1	79	3
201	61	BWoepk	60	26	14	8	7	4	FDPBs	26	30	28	21	20	2	519	CDFHPbs	28	21	16	14	4	0	77	1
202	79	BWypsc	32	30	12	8	4	3	PDFBs	25	34	23	23	18	3	978	HCpdfbs	45	34	7	9	8	8	68	2
203	79	BWyegs	38	33	19	6	3	3	PFDBs	25	34	26	20	15	2	673	CHpdfbs	36	34	9	8	7	1	76	2
204	86	BWesyo	40	32	9	5	4	4	PFBDs	17	37	22	20	19	3	486	CHpdfbs	59	12	7	7	5	0	92	3
205	89	BWepso	43	36	8	5	4	3	DFPbs	16	48	22	15	12	4	513	CHpfdbs	59	17	10	6	3	0	95	2
206	76	BWepoy	39	31	9	6	4	3	DFPBs	34	31	31	25	12	1	1019	HCfdpbs	44	25	11	10	6	4	93	2
207	76	BWGspe	33	28	18	6	6	4	DFPBs	25	29	26	22	21	2	726	CHdfpbs	39	29	10	10	5	0	75	0
208	68	WBYsfe	34	20	14	6	2	1	FPDBs	26	34	24	24	18	2	630	CHDfbps	33	29	15	9	5	5	68	1
209	73	WBGYpc	31	20	10	6	5	4	DPBFs	26	37	26	18	17	2	784	HCfdpbs	43	30	6	7	6	5	32	1
210	68	BWycpo	37	31	5	5	4	1	DPBFs	25	27	26	23	22	2	604	CHDfpbs	33	33	7	7	7	4	61	0
211	54	BWoeyz	59	27	6	2	1	1	BDSpf	27	35	32	12	11	11	374	CDBfsph	35	23	12	9	7	1	81	1
212	29	BWoekm	47	40	9	2	1	1	BDfps	43	38	33	13	10	4	442	DBCFpsh	35	23	15	14	7	6	73	0
213	51	BWocke	56	24	8	2	2	2	DBPFs	37	36	21	18	18	9	521	DHCPFbs	26	22	20	10	10	5	43	1
213A	42	BWcokp	53	18	7	8	2	3	DBPFs	45	29	27	24	18	5	807	HDCPbfs	43	18	10	13	10	4	43	2
214	38	BWorem	45	31	12	7	2	1	DBFsp	39	39	28	12	11	9	433	DCBfsph	37	21	16	10	8	2	74	1
215	45	BWoezy	52	30	10	5	1	2	DBFps	32	35	27	16	12	9	412	DCFBpsh	28	27	14	14	6	7	31	1
216	52	BWoekp	56	29	7	2	2	1	DBPsf	28	35	32	13	10	9	376	CDBpsfh	36	26	16	7	7	6	91	2
218	55	BWoegp	54	29	6	2	2	2	DBpfs	26	36	30	9	9	5	369	CDBpsfh	36	14	14	10	4	7	98	1
219	52	BWoypk	58	29	4	5	2	2	DBPFs	29	51	25	12	8	5	426	DCBhpfs	36	31	11	7	6	4	103	1
220	51	BWoekp	57	27	5	2	2	2	DBPsf	39	28	28	20	19	6	494	CDBHPfs	41	24	14	11	11	6	82	1
221	57	BWogye	49	30	6	4	3	1	DBPfs	26	41	27	20	6	7	420	CDHBpsf	36	26	12	8	8	5	62	2
222	68	BWogey	55	27	4	3	2	3	DBPfs	23	36	26	22	10	5	449	CDHpbfs	37	20	18	10	4	3	75	1
224	60	BWokyr	62	26	5	5	3	1	DBPfs	26	49	23	15	15	5	368	CDbpfhs	25	23	16	16	8	4	99	4
225	45	BWocep	47	28	7	6	3	3	DFPBs	51	33	26	13	16	2	732	DHFCPbs	25	19	13	9	11	8	68	3
228	35	BWokec	59	25	9	2	1	1	DPFbs	46	53	17	15	13	3	607	DHCfpbs	42	19	13	9	8	7	57	2

241

Dis-trict	Percent-age under tillage	Crop combinations	Proportion of tillage under six leading crops in rank order						Live-stock units per 100 acres	Live-stock combin-ations	Proportion of livestock units under five leading livestock in rank order					Man-days per 100 acres	Enter-prise combina-tions	Proportion of man-days under seven enterprises in rank order							Crops and grass 000 acres	Rough grazing 000 acres
228A	38	BWCokp	47	19	14	6	3	3	48	DPBfs	36	29	18	15	2	1414	Hdpcbfs	67	13	6	5	4	4	1	49	4
229	37	BWcopy	50	22	6	6	5	2	48	DPFBs	40	20	19	18	3	765	HDCFpbs	30	26	14	11	9	8	2	54	3
230	37	BWoeyk	49	29	11	5	4	2	44	DBfs	53	17	14	14	2	564	DCHFbps	43	16	11	11	9	8	2	52	1
232	44	BWCohp	45	19	15	5	4	3	36	DFBPs	41	19	19	15	6	1079	HCDbfps	49	22	14	5	4	2	2	117	3
233	50	BWChpg	38	22	11	7	7	3	17	BSDPf	30	22	18	16	13	1392	HCdbsfp	62	29	2	2	1	1	1	74	3
234	39	BWohpy	48	22	9	7	2	2	36	DSBFp	30	25	18	15	12	932	CHDsfbp	34	34	12	7	5	4	3	80	2
235	40	BWgpoc	40	22	6	6	5	5	27	SBDFp	30	22	16	16	16	1080	HCsdbfp	62	19	6	4	3	3	3	104	4
236	68	BWPQXz	29	20	9	9	4	4	21	BDPSf	28	27	19	16	11	920	HCdbpsf	52	30	6	4	3	3	2	88	2
236A	50	BWogpy	45	24	9	4	4	2	30	DSBPf	28	21	20	16	14	836	HCdsbpf	45	25	11	6	5	4	2	65	3
237	48	WBYPoj	33	22	13	10	6	5	35	SBDFp	43	18	16	13	9	594	CHSDfbp	37	18	18	10	7	6	4	94	2
238	36	BWokce	62	22	5	3	2	1	57	DFbps	41	32	13	12	3	789	FDHCbps	32	31	12	11	7	6	2	46	2
239	27	BWOkvy	54	23	14	2	1	1	41	DBpfs	57	22	9	8	4	462	DHCBfps	43	18	14	11	7	5	3	62	7
240	27	BWOhpk	38	29	14	6	1	1	47	DSBpf	33	27	23	10	7	711	CDHSbpf	27	22	20	14	8	5	4	80	3
241	27	BWopck	62	20	10	4	2	1	45	DBfps	42	33	9	9	7	561	DHBCfps	36	18	16	12	8	5	5	67	4
242	43	BWokzr	61	25	4	2	1	1	35	DBfps	49	29	10	6	5	471	DHCBfsp	37	19	18	14	6	3	3	57	6
244	63	WBFoye	32	28	7	6	5	5	35	DFBps	49	17	16	14	4	983	HCDfbps	50	19	18	4	4	3	1	69	6
245	38	BWokpe	51	28	9	5	3	1	36	DBpfs	61	24	8	5	3	551	DHCbpfs	41	21	19	11	4	3	2	67	2
246	44	BWozek	47	28	8	4	2	2	38	DBfps	47	27	12	10	3	824	HDCbfps	45	23	13	9	6	4	1	71	5
247	62	BWoeky	65	22	3	3	3	2	31	DFBPs	37	21	20	17	6	468	CDFBphs	31	25	16	9	9	8	3	130	6
248	60	BWoyke	69	24	5	3	2	2	31	DBpfs	47	20	18	13	3	455	DCBpfhs	33	32	10	8	7	7	2	108	3
249	56	BWoker	54	25	4	4	4	2	37	DFBps	46	21	16	14	3	559	DCHfbps	31	23	21	9	7	6	2	96	4
250	54	BWopky	50	29	6	3	3	2	37	DBFps	48	21	17	11	3	591	CDHfbps	31	30	13	10	9	5	1	89	4
251	46	BWCZpo	33	22	12	10	7	3	50	DFBps	43	25	16	15	3	1164	HDCfbps	50	20	12	9	5	4	1	47	2
253	34	BWkzco	55	19	8	4	3	3	28	DFPbs	51	17	15	15	1	161	DHFCpbs	38	23	14	10	8	7	1	62	72
254	41	BWopkz	53	20	5	5	3	3	40	DBfs	61	21	12	4	2	2608	DHCbpfs	43	20	20	9	5	2	1	60	6
255	39	BWozec	54	31	5	2	2	1	37	DBfs	40	25	13	11	11	461	DCHBfps	31	19	16	13	8	7	7	75	4
256	53	BWzoex	52	25	12	5	2	1	24	DBSpf	33	32	18	9	8	446	CHDBsfp	28	26	19	11	8	4	4	109	2
257	62	BWokym	58	29	4	2	1	1	22	DBSpf	48	22	14	10	5	364	CDbspfh	43	31	9	8	5	3	2	81	1
258	30	BWopke	66	24	4	2	2	1	39	DBfps	49	27	10	9	5	423	DCBpfhs	47	16	16	6	6	5	4	71	5
259	34	BWopkc	60	28	5	2	2	1	41	DBFps	51	20	17	8	4	502	DCHBfps	44	17	12	11	6	5	4	77	3
260	18	BWkope	49	24	7	6	4	2	52	DBPSf	45	19	13	12	11	484	DHBCSpf	42	13	11	9	9	8	8	50	5
260A	27	BWopsk	47	29	8	5	2	2	51	DBFps	45	24	15	10	6	724	DHCBfbs	33	28	12	11	8	5	4	56	2
261	17	BWopkm	45	34	10	4	2	2	47	DBpfs	65	19	7	6	3	526	Dbchpfs	63	11	9	5	4	2	2	60	1
262	36	BWokpy	61	30	4	1	1	1	38	Dbfps	66	15	10	6	2	468	DCbfphs	58	18	8	7	4	3	2	114	3
263	61	BWoked	62	28	4	2	1	1	23	DBfps	58	19	10	9	4	346	DCbfhps	37	36	9	6	6	4	3	95	20
264	33	BWokpm	59	30	5	3	1	1	44	DBfps	59	16	15	10	1	540	DFCbphs	52	16	15	9	6		1	99	2

ID																									
265	32	BWokpc	59	29	3	3	2	50	DFbps	59	16	12	12	1	593	DCFbphs	54	13	12	7	7	6	1	95	6
266	42	BWkoer	63	28	4	4	1	36	DBFsp	57	20	9	7	6	453	DCBhfsp	47	20	10	8	6	6	1	93	9
267	56	BWoker	59	28	3	4	1	27	DBFps	48	23	14	5	4	385	DCBfshp	35	32	11	11	7	5	4	122	15
268	31	BWkoet	56	25	7	5	2	45	Dbpfs	67	14	12	5	2	519	Dcbpfhs	63	14	8	11	4	4	2	104	5
269	28	BWkopd	61	23	6	6	1	50	DFBps	65	15	11	7	2	541	Dcbpfsh	48	12	9	7	6	2	1	99	4
270	39	BWokpv	60	21	8	5	1	42	Dbfps	57	15	15	9	4	512	DCHbfps	58	16	11	8	8	2	3	110	15
271	35	BWOkqv	54	27	7	6	1	40	Dbpsf	68	18	9	4	4	500	DCbfhps	67	18	6	4	4	2	2	43	4
272	17	BWokpc	51	21	16	1	3	48	DBfps	68	18	7	5	4	520	Dbcpshf	67	11	5	4	4	2	2	93	7
273	8	BWokpc	55	19	3	3	1	56	DBfps	62	18	9	7	4	596	DHbfpsc	59	15	3	6	6	4	3	66	5
273A	15	BWOkmp	61	17	6	6	3	53	Dbpfs	60	20	10	8	2	585	DBhcfps	59	12	3	6	6	3	3	56	5
274	22	BWokve	56	26	7	5	3	50	Dbpfs	67	16	9	8	1	594	Dcbhfps	62	9	7	9	5	5	2	79	3
275	16	Bwokvm	54	25	9	5	4	55	Dbpfs	68	13	12	6	2	611	Dbpcfhs	69	7	7	6	4	4	1	81	2
275A	10	BWopke	63	15	7	5	1	60	DBpfs	68	13	11	9	4	674	DCHbpfs	68	7	7	7	1	1	3	71	1
276	23	BWkpot	48	28	8	6	2	49	DBpfs	58	18	12	9	3	577	DCHfbps	51	12	10	8	6	6	3	75	2
276A	20	BWokpt	46	24	8	7	2	58	BSDpf	61	13	12	11	8	684	DHCbpfs	55	10	9	6	7	7	3	60	2
277	26	BTWokr	53	17	7	6	4	53	DBfps	55	18	10	9	2	650	SBDChpf	46	14	13	6	6	6	2	67	2
278	16	BWopkt	49	15	15	7	2	30	Dbpfs	38	38	19	3	6	221	DCBFhps	31	25	19	15	3	2	5	96	39
279	26	BWokve	51	24	7	5	4	52	SBDpf	70	13	10	11	6	562	Dbpfschs	45	13	12	11	8	6	3	69	3
279A	7	BOTkqr	42	39	7	4	3	55	DSBpf	51	21	11	10	5	576	SDBchpf	74	8	6	8	7	3	1	60	2
280	14	BOMkwt	42	17	11	6	2	39	DBSpf	39	35	23	5	2	293	DSBchpf	31	25	23	12	9	1	1	80	20
280A	15	BOmkwt	49	15	10	8	4	52	DBSpf	44	23	21	9	3	505	DSBCpfh	44	18	13	9	4	6	2	12	1
280B	24	BMokrp	54	14	8	6	6	46	DBfsp	37	27	26	5	4	452	DBScpfh	39	20	17	16	6	3	2	50	4
281	22	BOmkrw	54	18	11	7	5	50	DBspf	49	22	15	9	6	472	DBfscph	51	14	12	11	7	2	2	62	10
282	13	BOmkrw	59	15	14	5	2	51	DBSfp	47	26	13	13	8	468	DBScfph	54	18	9	7	7	1	5	52	9
282A	14	BOmwkt	60	14	10	8	4	45	SDBfp	44	33	15	13	1	419	DBScfph	49	18	7	7	5	0	5	55	6
283	13	BOMtkw	53	19	9	6	3	40	DBFsp	37	33	15	15	7	273	DSBCfph	43	24	13	8	5	5	1	57	19
284	18	Boktwm	54	13	10	5	2	41	DBSfp	34	30	29	17	3	358	DBFScph	32	27	20	8	3	1	1	77	15
285	17	BWOtkm	59	9	8	8	6	57	DFBps	42	22	17	17	8	563	DBCSfph	45	14	14	13	5	5	2	59	2
285A	23	BWokmt	51	14	13	8	6	49	DFBps	45	23	14	10	8	514	DFBpchs	45	15	14	10	3	3	2	62	4
286	17	BWOmkp	55	13	10	7	6	62	DBFps	52	17	17	14	4	626	DFpbchs	53	13	13	11	9	5	3	44	4
286A	19	BWOmkp	55	11	11	6	3	73	DFBps	49	19	14	14	4	750	DHCBfps	48	18	9	7	7	5	3	49	2
287	30	BWkopt	48	17	7	7	4	55	DBSpf	45	19	15	11	6	720	DHCBspf	36	19	14	9	8	5	3	58	1
287A	28	BWOktp	51	10	7	7	5	40	BDspf	45	19	14	13	9	539	DHCBspf	29	20	16	11	7	5	3	47	3
288	19	BKWToz	45	11	9	7	6	36	DBSpf	36	30	13	13	6	297	DBHCspf	33	21	17	12	9	6	5	55	19
288A	24	BKowtm	55	11	11	7	4	41	BDSfp	38	36	13	7	9	341	DBCShpf	41	16	14	12	9	4	4	66	13
289	15	BKOwtp	53	12	9	6	5	19	DBSfp	44	29	15	5	4	71	DBHScfp	28	25	16	12	4	4	3	58	135
289A	22	BOWmtk	53	13	11	8	5	43	DBSpf	43	28	14	9	5	329	DBSCfp	35	20	13	16	4	6	2	65	15
290	28	BKowtm	56	9	7	8	5	49	Dbpsf	32	31	14	8	7	537	DCBSphf	43	16	14	7	6	5	4	79	4
290A	13	BWkotm	62	12	10	6	6	54	DBspf	46	26	10	5	9	561	Dbpcsth	67	10	15	5	5	5	2	51	4
291	19	BMorkw	63	16	11	8	2	49	DBPsf	65	15	14	8	7	459	DBScpfh	47	21	10	6	4	4	0	62	6
291A	14	BMopkr	61	13	13	8	3	51	BDSpf	43	32	14	8	4	349	DBSfcph	37	28	12	8	7	4	7	60	15
292	24	BPOkmq	55	11	11	4	5	52	DBPsf	39	31	13	10	9	711	DHBCpsf	29	27	8	13	8	7	4	52	2
293	19	BMOpkr	54	14	11	5	4	47	BDSpf	38	33	14	14	8	410	DBCSphf	34	25	14	12	6	4	3	54	10
293A	22	BMopkw	39	28	10	6	6	48	DBspf	40	35	10	9	6	491	DBCspfh	40	21	10	9	4	4	4	50	4

District	Percentage under tillage	Crop combinations	Proportion of tillage under six leading crops in rank order	Livestock units per 100 acres	Livestock combinations	Proportion of livestock units under five leading livestock in rank order	Man-days per 100 acres	Enterprise combinations	Proportion of man-days under seven enterprises in rank order	Crops and grass 000 acres	Rough grazing 000 acres
294	32	BMkowr	49 33 5 3 3 2	40	BDspf	40 33 12 10 4	372	DBCspfh	33 23 20 10 7 5 3	52	7
295	18	BMowrp	44 29 9 6 3 3	46	BDspf	42 30 14 10 4	340	DBScpfh	34 30 13 11 8 3 1	58	14
296	31	MBokwp	38 34 6 6 5 4	46	DBpsf	41 34 14 8 2	491	DBCpshf	40 20 19 9 8 2 1	73	6
296A	24	MBQkpo	32 29 8 6 5 4	58	DBpfs	49 25 15 7 4	744	DHBCpfs	41 21 12 11 8 4 3	43	2
297	26	QMBcop	27 22 22 8 4 4	64	DBpfs	51 24 14 10 1	798	DHBcpfs	44 19 12 10 8 5 1	41	2
298	22	MBQokp	33 25 17 7 5 4	57	DBpfs	60 22 13 3 1	630	DBcphfs	57 13 10 8 8 2 1	43	7
299	29	QMBPck	33 16 15 11 8 5	55	Dbpfs	69 18 10 3 0	729	DHcbpfs	48 26 12 8 5 1 0	44	11
300	12	BOwpkt	55 19 15 5 2 1	45	BSDpf	49 22 21 4 3	372	BDScfph	34 27 21 9 4 3 1	38	5
301	10	BOkmrp	60 25 4 4 3 1	45	BDSpf	52 23 21 3 1	323	BDScpfh	36 28 24 7 3 1 0	40	6
302	16	BOWpkm	55 15 13 8 3 2	43	BDSpf	53 22 19 4 2	361	BDScphf	35 25 18 14 4 3 2	35	6
302A	7	BOwkpr	55 22 8 5 3 3	56	BSDpf	48 21 18 11 1	411	BDSpcfh	36 25 23 10 4 2 0	14	1
303	13	BOMpwv	46 24 13 6 3 3	48	BDSpf	40 29 25 3 3	372	BDScpfh	35 27 23 9 3 3 0	29	8
304	10	BOpmvw	48 23 8 8 3 3	43	BDSpf	44 26 24 4 3	351	BDScfhp	30 30 22 7 5 4 3	28	17
305	6	BOMkrp	42 20 9 6 6 5	33	SBdfp	41 38 15 4 2	199	SBDfhcp	40 31 20 4 3 2 1	22	34
306	7	BORPmt	41 18 10 9 7 4	24	SBdfp	63 20 11 4 2	118	SBDfhcp	56 15 14 5 4 3 2	23	76
307	7	BOkvpw	57 12 7 5 5 3	24	SBDfp	47 28 17 5 4	121	SBDfpch	44 22 21 5 4 3 2	19	51
308	7	BORtvm	27 27 21 5 5 5	45	SBDpf	48 29 20 2 2	362	SDBchpf	43 25 22 5 2 2 2	59	25
309	11	BORmwt	44 14 14 9 6 4	46	SDBfp	40 26 24 6 2	273	SDBcfph	34 31 17 7 6 3 1	46	22
309A	19	Bowmkr	60 11 9 8 4 3	57	SBDfp	44 24 20 7 5	495	SDBcfph	48 17 15 10 5 4 1	31	3
310	9	BROMkw	37 17 16 16 6 5	38	DBSpf	41 28 26 3 2	283	SDBcfph	37 32 21 5 2 2 1	61	35
311	13	BMWopc	41 20 13 10 5 2	53	DBSpf	49 23 15 7 6	522	DBSchpf	52 15 12 7 5 5 4	50	10
312	18	Bowmpk	62 11 9 7 3 2	56	DBsfp	44 25 19 9 3	532	DSBchpf	44 16 15 9 6 4 4	41	4
313	19	BMPowx	53 12 11 9 5 2	53	DBpfs	53 22 11 9 5	635	DHFBcsp	42 17 11 11 9 7 3	38	4
314	12	BMwokp	50 21 10 8 4 4	70	SBdfp	59 17 11 7 5	696	Dbpfcsh	64 11 7 6 5 1 0	33	1
315	3	RBOmvt	33 28 16 9 5 4	24	SBdfp	59 26 14 1 1	119	SBdcfph	56 21 18 4 1 1 0	44	84
316	3	ROVPbv	37 18 18 16 6 3	22	SBdfp	62 28 8 1 1	135	SBdhcfp	60 23 11 2 2 1 1	23	70
316A	3	ROvbpt	54 22 9 7 4 2	26	SDBfp	68 26 4 0 0	170	Sbdcfph	68 22 6 1 1 1 0	29	77
317	7	BOMRwv	37 21 14 13 9 3	40	SBdfp	34 27 27 8 5	330	DSBfpch	34 31 20 8 4 0 0	48	32
317A	9	ORBTwm	31 26 18 11 5 5	34	DSBpf	53 33 10 4 3	257	SBdcfph	48 28 14 7 2 2 0	56	41
318	8	BOWMtv	37 18 16 15 8 3	40	DBSfp	33 31 29 4 3	327	DSBfcph	38 27 21 5 3 0 0	51	27
318A	16	BWOMtr	45 17 15 13 4 3	57	SBDpf	40 26 17 11 6	532	DBSfcph	45 18 14 10 9 4 1	21	1
319	6	BORMwv	31 28 14 10 7 3	40	SBdfp	39 37 21 3 2	267	SBDcpfh	35 30 28 4 3 1 –	42	19
320	7	ROBMtw	28 24 23 13 5 4	32	SBfdp	52 28 16 2 2	236	SBDcfph	49 22 21 4 2 1 –	49	50
321	13	ORBtwp	35 23 20 11 5 2	38	SBcdp	55 41 2 1 1	229	SBcdfph	51 34 8 4 2 0 0	63	23
322	9	ORtbvp	42 34 13 4 4 2	22	SBcdp	62 33 3 1 1	95	SBcdpfh	57 28 8 4 1 1 –	44	63
323	12	OBRTwp	31 21 19 13 7 2	35	SBdpf	56 39 3 1 1	160	SBcdpfh	51 32 11 4 1 1 0	55	31

244

No.	n	Code								Code						Total	Code									
324	8	OBRmkp	37	23	20	5	5	3	32	SBDfp	50	25	22	1	1	254	SDBcfph	46	28	19	5	2	1	0	58	51
325	8	ORbkmp	38	36	14	4	3	3	22	SBDpf	53	23	21	1	1	156	SDbcpfh	49	28	18	4	1	1	1	30	63
326	8	BOMrpk	34	29	17	9	4	3	40	DBSpf	51	27	20	2	1	380	DBScpfh	57	18	17	6	2	2	—	48	9
326A	14	BMOrkp	41	24	19	7	4	3	42	DBspf	54	27	16	2	1	429	DBscpfh	57	17	14	9	2	1	0	57	6
327	16	BMOrkp	52	20	18	4	3	2	49	DBspf	66	23	4	4	1	517	Dbcspfh	68	15	9	8	3	1	0	47	2
328	9	ORTkbv	37	34	21	2	2	1	38	SBdfp	67	25	6	1	1	283	SBdcfph	62	21	8	7	1	1	0	37	32
329	11	OTRbpw	40	30	18	7	1	1	38	SBdpf	61	32	6	1	1	301	SBcdhfp	54	25	11	8	1	1	0	60	41
330	14	BOTWrp	29	24	18	12	5	4	48	SBdfp	50	36	12	2	1	409	SBCdphf	42	26	14	14	2	1	1	66	12
332	15	BWOTKP	25	20	17	10	9	6	42	DBSfp	32	32	27	6	3	329	DSBCHfp	31	20	19	11	10	1	1	47	11
333	9	BWOptk	40	26	12	5	4	4	44	DBSfp	39	32	23	3	3	440	DBScf	40	21	18	12	4	2	7	52	2
334	11	BCWOUP	24	22	14	8	8	7	29	SDBpf	36	30	24	5	5	160	DSBHcfp	30	29	14	13	4	4	2	33	28
335	13	BWOkpt	36	25	10	6	6	3	44	DBsfp	49	31	13	4	2	458	DBSchfp	48	19	11	8	5	2	4	46	2
336	20	BWOKpt	36	23	11	9	7	4	50	DBSfp	48	18	15	13	5	563	DFSBHcp	41	17	11	10	8	3	4	17	1
338	6	BPOMCt	34	17	12	8	5	5	33	DBSfp	35	31	18	12	2	235	DCBSHfp	32	18	17	13	9	2	5	35	20
340	16	BROPMt	33	18	9	8	6	6	26	SBDpf	38	36	19	10	4	159	SBDcphf	37	27	24	5	11	2	2	34	40
342	6	BWOpmr	58	15	11	5	2	2	41	DBSfp	37	35	19	5	4	335	DBSchp	40	22	16	11	3	4	7	52	12
343	15	BRWCPk	32	18	13	13	6	6	26	SBDfp	46	25	20	5	4	148	SDBfchp	42	25	18	5	4	4	2	34	49
343A	6	BWocpu	54	20	10	4	3	2	47	DBspf	45	31	10	8	7	481	DBHscfp	45	18	10	8	8	4	4	32	3
344	3	BMOvpk	49	19	18	5	4	2	40	Dbspf	67	22	5	4	1	306	Dbscpfh	73	15	5	4	3	5	3	55	12
344A	6	BMOkpr	49	25	8	6	4	4	48	Dbspf	71	22	5	1	1	483	Dbscpfh	76	14	4	4	1	0	0	46	3
345	2	BORVkm	29	22	18	4	5	5	39	DSBfp	52	25	21	5	1	281	DSbcfhp	58	23	15	2	1	0	0	50	15
345A	2	BOMvkp	36	28	18	6	4	4	41	DBspf	65	22	10	1	2	380	Dbscpfh	69	15	9	2	1	1	1	50	7
346	6	ROTkvp	38	22	18	6	5	5	39	DSBfp	41	34	23	2	1	241	DSbcfph	49	31	16	2	1	1	1	43	17
346A	2	Rokbtm	67	9	7	7	4	3	29	SDBfp	46	31	22	1	1	158	SDbcfph	43	39	16	1	1	0	0	52	43
347	13	BMOPkc	49	12	9	8	5	4	67	DFbsp	47	38	13	1	1	638	DFbchsp	47	39	8	3	1	1	1	39	7
347A	16	MWCOKU	40	17	12	9	8	9	74	DFbsp	45	43	10	1	1	665	DFbsphc	56	33	8	1	1	0	1	17	3
347B	27	BOMKpr	20	16	12	11	8	5	36	Dbsfp	70	21	5	2	2	328	Dbspfch	76	14	5	2	2	1	2	19	5
348	10	BMorpv	39	35	15	5	3	2	45	DBspf	60	23	13	4	1	360	Dbscpfh	63	14	11	8	3	1	2	41	10
348A	22	BMorkp	52	22	13	7	3	2	43	DBspf	57	22	16	4	1	345	DBscpfh	61	14	13	8	3	1	0	43	15
349	28	BMOork	39	27	13	9	4	3	42	DBpsf	52	39	4	3	3	458	DCBpshf	49	22	21	3	2	1	1	51	11
350	10	BOmpkv	55	18	13	11	3	3	46	DBsfp	65	28	3	2	2	477	Dbcsfph	71	18	6	7	1	1	1	47	6
351	22	BPOmik	43	22	14	11	2	2	36	DBsfp	54	33	6	4	3	348	DCBhsfp	46	22	17	4	4	0	2	46	13
352	28	BPOMqk	34	29	15	14	3	1	41	BDspf	45	44	5	4	2	563	DCBhfsp	34	31	20	6	4	3	2	48	5

245

TABLE II

In this table details are given for the ten leading districts for each of the choropleth maps. Owing to rounding, figures given here may not agree exactly with those used in constructing the maps.

Figure 13

District	Per 100 acres
319	79
175	79
327	75
291	75
282	75
346	74
317A	74
344	74
323	73
324	73

Figure 14

District	Per 100 acres
52	75
61	73
300	73
305	70
302	69
303	69
302A	69
347B	68
334	67
301	66

Figure 24

District	Per 1,000 acres
228A	73
189	64
155	62
251	59
233	57
161	57
154	56
153	56
156	52
246	50

Figure 25

District	Per 1,000 acres
69	14
251	14
155	14
71	14
53	13
297	13
296A	12
189	12
299	12
80A	12

Figure 26

District	Per 100
347B	70
80	70
46	70
319	69
25A	69
346	69
80A	68
43	68
328	68
81	68

Figure 27

District	Per 100
345A	79
347B	75
325	74
322	73
346A	73
345	71
328	71
326A	69
327	67
80	67

Figure 28

District	Per 1,000 acres
228A	52
246	38
155	38
154	34
153	33
189	32
156	32
200	30
233	30
161	29

Figure 29

District	Per 1,000 acres
233	17
251	17
189	15
236	14
129	13
161	12
235	12
190	12
352	11
153	11

Figure 46

District	Per 100 acres
306	77
3	76
316	75
307	73
34	73
316A	73
289	70
25A	68
325	68
43	66

Figure 47

District	Per 100 acres
253	93
289	82
38	80
239	76
194	70
288A	70
105	70
39	70
260	70
279	69

Figure 48

District	Per 100 acres
156	95
157	94
190	94
153	93
187	93
155	93
154	91
129	90
161	89
189	89

Figure 49

District	Per 100 acres
212	93
263	93
266	92
258	92
262	92
264	92
211	92
201	92
214	92
148	92

Figure 51

District	Per 100 acres
190	32
194	31
157	31
209	30
191	29
185	29
137	29
204	29
195	29
156	28

Figure 52

District	Per 100 acres
209	41
114	39
208	39
237	39
279A	38
212	38
113	38
194	37
148	36
214	36

Figure 53

District	Per 100 acres
154	74
153	70
156	69
155	68
190	66
187	65
189	60
191	60
209	60
161	58

Selected Statistics

Figure 54

District	Per 100 acres
26	48
158	44
175	44
167	42
199A	42
176	42
168	42
174	42
164	42
169	42

Figure 55

District	Per 100 acres
48	71
7	69
35	69
34A	68
34	67
26	67
35A	65
247	64
258	64
40	64

Figure 56

District	Per 100 acres
52	96
48	91
53	85
116A	83
34	83
25A	82
46	82
68	81
35	80
61	79

Figure 57

District	Per 100 acres
23	6
1	6
112	6
32	5
92	5
95	5
143	5
93	5
22	5
91B	5

Figure 58

District	Per 100 acres
322	41
329	40
325	38
324	36
321	35
328	34
323	31
317A	30
326	29
319	27

Figure 59

District	Per 100 acres
328	95
322	89
43	88
346	85
329	82
316	75
316A	75
325	69
321	56
324	56

Figure 60

District	Per 100 acres
296	12
349	11
294	10
298	8
296A	8
297	7
293A	6
299	6
295	5
348	5

Figure 61

District	Per 100 acres
349	40
296	39
298	37
348	34
296A	34
294	33
295	29
293A	28
297	28
344A	25

Figure 62

District	Per 100 acres
44	71
349	52
298	51
296A	47
296	46
299	44
297	44
348	39
294	37
347A	37

Figure 63

District	Per 1,000 acres
159	24
178	23
169	13
177	10
174	5
129	5
270	4
263	3
164	3
192	3

Figure 64

District	Per 1,000 acres
178	31
159	31
18A	22
169	17
177	13
295	12
270	9
17	9
336	8
286	7

Figure 65

District	Per 1,000 acres
244	4
203	4
175	3
207	3
202	2
208	2
173	2
214	1
177	1
40	1

Figure 66

District	Per 1,000 acres
244	7
19	5
203	5
288A	4
214	4
207	3
175	3
208	3
202	3
228A	3

Figures 67

District	Per 100 acres
190	35
187	34
160	25
189	24
156	24
191	24
154	24
195	23
153	23
155	23

Figure 69

District	Per 100 acres
190	18
187	14
154	14
156	14
189	14
161	13
153	12
191	12
157	12
155	11

Figure 70

District	Per 100 acres
352	29
351	22
78	19
190	19
338	18
79	16
189	16
187	16
154	15
80	15

Figure 71

District	Per 100 acres
80	100
316	98
251	98
228	98
348	97
232	96
352	96
213A	96
233	96
237	95

Figure 72

District	Per 1,000 acres
352	66
154	58
351	41
236	39
299	37
78	33
202	30
349	23
153	23
132	21

Figure 73

District	Per 100 acres
299	91
351	86
352	81
297	75
316	70
300	67
296A	67
298	66
349	63
344A	57

Figure 74

District	Per 100 acres
187	19
190	18
188	15
163	14
160	14
164	12
195	12
158	12
162	12
178	12

Selected Statistics

Figure 75		Figure 76		Figure 77		Figure 78		Figure 79	
District	Per 100 acres	District	Per 100 acres	District	Per 1,000 acres	District	Per 100 acres	District	Per 100 acres
187	21	175	92	26	42	43	43	44	100
89A	20	181	92	1	35	329	30	3	97
190	19	169	91	27	35	45	21	329	93
188	18	180	89	329	33	328	20	4	91
163	17	167	88	2	31	3	17	328	90
89	16	164	86	15A	30	330	17	322	89
165	16	158	84	39	30	346	16	45	87
160	16	159	83	40	27	19A	15	278	86
178	16	174	82	33	25	278	14	317A	86
164	15	188	82	330	24	44	14	19A	86

Figure 80		Figure 81		Figure 82		Figure 83		Figure 84	
District	Per 1,000 acres	District	Per 1,000 acres	District	Per 100 acres	District	Per 1,000 acres	District	Per 1,000 acres
164	6	330	25	279A	28	102	51	44	60
167	6	323	19	294	19	154	50	347A	39
166	6	288	19	276	18	233	41	51	22
165	5	19	18	288	16	210	39	80	14
41	5	338	17	283	15	153	34	299	11
173	5	276	13	46	14	232	32	302A	11
36	5	328	12	272	14	299	32	102	10
36A	5	68	12	286	14	228A	28	346	10
168	5	280	12	19	13	208	27	52	9
49	5	17	12	298	13	236	26	43	8

Figure 85		Figure 86		Figure 87		Figure 88		Figure 89	
District	Per 1,000 acres	District	Per 100 acres	District	Per 1,000 acres	District	Per 1,000 acres	District	Per 100 acres
270	30	45	51	299	7	297	24	204	9
253	30	51	27	297	6	299	23	180	8
290	27	116	17	298	4	298	18	205	8
288A	27	52	17	296A	4	296A	15	185	7
244	25	46	14	293	3	347B	15	206	6
271	22	289	12	296	2	293	13	193	5
288	22	288	12	280B	2	68	13	194	5
287	21	288A	11	290	2	81	13	186	5
287A	21	347B	11	294	2	116	12	179	5
249	20	116A	10	293A	2	289	11	203	5

Figure 90		Figure 91		Figure 92		Figure 93		Figure 94	
District	Per 100 acres	District	Per 1,000 acres	District	Per 100 acres	District	Per 1,000 acres	District	Per 1,000 acres
204	10	322	31	346A	63	249	9	316A	32
180	10	325	29	316A	50	222	7	346A	29
185	9	321	28	316	35	267	6	249	16
205	9	328	27	346	35	224	5	214	14
206	8	317A	22	325	35	214	5	267	10
193	7	323	22	322	33	247	5	325	10
194	6	315	22	315	33	147	4	222	10
203	6	320	19	116A	32	211	4	224	9
186	6	329	19	328	31	248	4	16	8
179	6	324	17	320	27	221	3	147	8

Selected Statistics

Figure 95

District	Per 1,000 acres
156	10
186	9
159	9
192	9
249	9
188	8
184	8
248	8
178	7
222	6

Figure 96

District	Per 1,000 acres
80	69
316	16
345	14
81	13
316A	9
45	7
307	5
315	5
344	5
308	5

Figure 97

District	Per 1,000 acres
195	9
203	6
178	6
155	6
160	6
121	6
208	6
189	5
187	5
180	5

Figure 103

District	Per 100 acres
80	100
43	100
44	100
45	100
51	100
52	99
116A	99
347A	98
68	98
80A	98

Figure 104

District	Per 100 acres
263	59
294	56
281	55
247	54
78	54
248	54
296	52
26	52
270	52
178	52

Figure 105

District	Per 100 acres
262	48
265	47
261	47
263	47
275	47
275A	46
264	46
259	46
276	46
269	45

Figure 106

District	Per 100 acres
43	99
44	99
45	99
80	99
51	98
52	97
68	94
70	93
81	93
116A	93

Figure 107

District	Per 100 acres
43	99
44	99
45	99
80	99
51	99
52	98
70	97
35	96
68	96
81	96

Figure 108

District	Per 100 acres
262	45
261	45
265	45
275	44
275A	44
276	43
269	42
259	41
273A	40
279A	40

Figure 109

District	Per 100 acres
281	43
294	39
282	38
15A	37
293A	37
291	36
296	36
298	35
295	35
293	33

Figure 110

District	Per 100 acres
80	97
347B	86
346A	84
43	84
346	83
347A	81
44	81
116A	79
316	79
79A	79

Figure 111

District	Per 100 acres
43	100
44	100
45	100
80	84
81	83
19	80
51	77
52	77
18A	77
25A	76

Figure 112

District	Per 100 acres
184	66
161	59
195	59
189	58
27	54
33	52
199A	52
199	50
30	48
180	48

Figure 113

District	Per 100 acres
315	46
52	46
316	45
308	43
116A	42
4	41
309	41
80A	40
303	39
188	39

Figure 114

District	Per 1,000 acres
178	52
208	25
158	19
159	18
169	17
168	16
320	9
125	9
328	9
222	7

Figure 115

District	Per 1,000 acres
116A	65
178	60
320	43
316A	41
347A	39
208	31
328	28
325	22
159	21
158	21

Figure 116

District	Per 100 acres
154	37
153	31
233	27
161	27
156	27
155	25
189	23
197	22
235	22
102	18

Figure 117

District	Per 100 acres
154	34
153	27
197	20
156	19
155	17
161	16
135	14
157	13
129	12
137A	12

Figure 118

District	Per 100 acres
154	37
52	31
153	29
197	25
102	21
156	20
135	20
51	18
53	18
155	18

Figure 119

District	Per 100 acres
282	94
137A	94
131	90
33	89
165	88
173	88
130	86
137	86
163	85
125	82

Selected Statistics

Figure 120 District	Per 100 acres	Figure 121 District	Per 100 acres	Figure 122 District	Per 100 acres	Figure 123 District	Per 100 acres	Figure 124 District	Per 100 acres
197	12	131	9	71	7	71	8	154	18
196	8	130	5	154	6	154	6	153	17
102	6	71	5	153	5	197	6	135	9
71	5	135	5	232	5	153	6	71	8
153	4	137	4	135	5	135	4	236	7
256	3	154	4	197	4	232	4	155	4
154	3	153	4	236	3	251	2	102	2
135	2	196	3	233	2	236	2	132	1
194	2	163	3	251	2	84	2	84	1
236	2	173	3	158	2	132	2	156	1

Figure 125 District	Per 100 acres	Figure 126 District	Per 100 acres	Figure 127 District	Per 100 acres	Figure 128 District	Per 100 acres	Figure 129 District	Per 100 acres
236	14	160	13	56	11	131	7	156	6
299	10	158	12	50	9	173	5	161	4
153	7	178	10	59	4	165	5	157	4
297	7	71	8	102	3	130	5	189	4
154	6	159	5	210	3	163	4	155	4
135	4	191	5	65	3	59	3	208	3
298	4	187	4	106	3	134	3	153	3
71	2	169	3	233	3	129	3	202	3
296A	2	164	3	207	2	56	3	160	3
290	2	129	3	204	2	125	3	203	2

Figure 130 District	Per 100 acres	Figure 131 District	Per 100 acres	Figure 132 District	Per 100 acres	Figure 133 District	Per 100 acres	Figure 134 District	Per 100 acres
158	15	129	19	71	6	163	11	102	8
178	8	197	7	287	6	135	6	279	6
160	7	71	6	278	5	165	6	156	5
197	6	121	6	287A	4	244	6	155	3
159	6	59	5	285A	3	59	5	197	3
168	5	56	4	277	3	175	5	277	3
84	4	187	4	286A	3	154	4	251	3
164	4	160	3	285	2	161	4	208	2
187	3	232	2	288A	2	164	3	232	2
186	3	102	2	237	2	133	3	71	2

Figure 135 District	Per 100 acres	Figure 136 District	Per 100 acres	Figure 137 District	Per 100 acres	Figure 138 District	Per 100 acres	Figure 139 District	Per 100 acres
71	12	197	9	180	11	102	21	160	14
59	8	232	7	168	10	232	8	153	12
237	7	235	6	156	9	197	5	187	11
161	7	233	6	190	8	228A	4	155	7
129	5	102	5	154	7	71	3	154	5
164	4	106	5	165	7	186	3	156	4
102	4	236	5	163	7	210	3	178	3
156	4	196	4	161	6	236	3	195	3
189	3	203	3	65	5	121	2	191	2
128	3	205	2	131	4	213A	2	189	2

Selected Statistics

Figure 140		Figure 141		Figure 142		Figure 143		Figure 144	
District	Per 100 acres	District	Per 100 acres	District	Per 100 acres	District	Per 100 acres	District	Per 100 acres
187	38	71	14	228A	13	155	20	200	7
160	18	197	5	249	4	156	18	206	5
71	10	228A	4	244	3	153	11	30	5
184	6	160	4	189	3	154	6	71	4
154	3	232	3	124	2	157	4	244	4
128	2	229	3	161	2	189	4	246	3
190	2	102	2	207	2	161	3	69	3
191	2	251	2	313	2	163	2	228A	3
135	1	84	2	117	2	292	2	102	2
129	1	213A	2	82	2	299	2	155	2

Figure 145		Figure 146		Figure 147		Figure 148		Figure 149	
District	Per 100 acres	District	Per 100 acres	District	Per 100 acres	District	Per 1,000 acres	District	Per 100 acres
197	6	197	8	131	5	189	38	233	5
156	4	71	6	165	4	161	36	273	4
191	4	196	5	173	4	233	24	189	4
299	3	102	4	163	4	235	15	161	4
187	3	158	4	130	4	251	13	235	4
155	3	160	3	59	4	85	12	251	3
178	2	178	3	129	3	156	11	260A	3
188	2	256	2	56	3	234	11	104	3
185	2	153	2	134	3	163	11	95	3
184	2	56	2	156	2	104	9	234	3

Figure 150		Figure 151		Figure 152		Figure 153		Figure 154	
District	Per 100 acres	District	Per 100 acres	District	Per 100 acres	District	Per 100 acres	District	Per 100 acres
161	10	163	6	189	10	233	14	233	20
189	8	95	5	233	8	235	7	235	17
233	7	85	4	161	7	232	6	236A	10
235	6	92	4	195	6	163	4	232	9
160	5	104	4	235	5	239	4	234	7
156	4	98	4	232	4	105	4	102	6
251	4	162	3	85	4	236A	3	189	6
234	3	96	3	104	3	234	2	236	6
203	2	202	3	163	3	240	2	104	6
232	2	165	3	154	2	165	2	161	6

Figure 155		Figure 156		Figure 157		Figure 158		Figure 159	
District	Per 100 acres	District	Per 100 acres	District	Per 100 acres	District	Per 100 acres	District	Per 100 acres
235	13	104	6	233	14	233	6	233	8
233	11	98	4	235	9	236A	6	235	4
232	7	102	3	236A	7	235	4	189	3
236A	5	105	3	232	7	202	4	232	3
236	4	258	3	234	6	236	4	161	3
234	4	235	3	236	5	232	4	234	2
102	3	260	2	202	5	234	3	236A	2
202	3	95	2	189	5	207	2	240	1
189	2	260A	2	161	4	97	2	236	1
161	2	93	2	96	4	189	2	104	1

Selected Statistics

Figure 160		Figure 161		Figure 162		Figure 163		Figure 164	
District	Per 1,000 acres	District	Per 1,000 acres	District	Per 1,000 acres	District	Per 1,000 acres	District	Per 1,000 acres
96	30	235	30	277	4	102	44	235	33
98	18	233	29	275	2	195	24	233	13
95	14	236A	14	260A	1	235	19	236A	8
104	13	232	10	279	1	233	14	105	4
92	13	189	10	98	1	260	12	104	3
93	10	236	8	258	1	161	8	234	3
287	8	234	6	276	–	105	8	232	3
276	8	161	6	260	–	189	7	236	3
286	7	202	4	95	–	191	7	222	2
275A	7	103	4	96	–	104	7	213	1

Figure 165		Figure 166		Figure 167		Figure 168		Figure 169	
District	Per 100 acres	District	Per 100 acres	District	Per 100 acres	District	Per 100 acres	District	Per 100
98	3	98	8	69	106	69	73	344A	93
233	3	234	7	347A	74	13	70	350	93
234	3	233	7	286A	73	77	70	349	91
105	2	240	6	77	71	14	69	347B	91
232	2	105	6	314	70	314	68	81	90
240	2	95	5	347	67	79B	66	327	90
95	1	232	4	79B	65	76A	66	344	90
235	1	96	3	76A	65	82A	66	80	89
96	1	235	3	53	65	80A	65	352	89
236A	1	104	3	76	65	297	64	120	88

Figure 170		Figure 171		Figure 172		Figure 173		Figure 174	
District	Per 100	District	Per 100 acres	District	Per 100 acres	District	Per 100	District	Per 100
79B	72	77	54	77	40	79	78	299	97
344A	71	69	54	79B	40	279A	77	76	97
81	70	79B	51	79	38	275A	77	228	97
279A	70	80A	47	80A	36	79B	77	77	97
347B	70	79	47	76A	35	275	77	79B	96
299	69	76A	46	69	34	77	77	87	96
120	69	314	46	314	34	265	76	269	96
79	69	275A	46	299	34	76	75	88	96
77	69	82A	45	275A	33	268	75	276A	95
275	68	120	44	81	32	276A	74	78	95

Figure 175		Figure 176		Figure 177		Figure 178		Figure 179	
District	Per 100 acres	District	Per 100	District	Per 100 acres	District	Per 1,000 acres	District	Per 100
69	8	156	100	302A	44	321	108	321	96
13	7	43	43	301	38	330	100	3	94
14	6	188	39	302	38	323	97	323	94
120	6	19	37	300	37	90	97	189	91
286A	6	1	37	295	36	319	85	322	90
299	6	18A	36	291A	35	329	85	154	87
19	6	68	36	90	35	295	81	1	85
88	6	263	35	303	34	291A	74	2	84
298	6	169	34	92	34	92	72	329	83
82A	5	178	33	293	33	317A	71	316A	82

Selected Statistics

Figure 180		Figure 181		Figure 182		Figure 183		Figure 184	
District	Per 1,000 Acres	District	Per 100	District	Per 1,000 acres	District	Per 100	District	Per 1,000 acres
33	105	189	616	302	91	156	47	302A	114
13	94	157	353	302A	88	161	32	301	104
14	92	33	245	300	80	190	28	302	86
36	89	161	240	303	75	155	26	141	73
295	87	27	180	13	69	153	26	300	72
47	87	56	172	352	68	157	24	147	68
349	87	190	169	301	68	185	24	146	62
95	85	197	163	296	65	56	23	142	58
36A	83	36	163	293A	65	189	23	145	54
90	81	138	162	212	64	184	22	352	45

Figure 185		Figure 186		Figure 187		Figure 188		Figure 189	
District	Per 100	District	Per 1,000 acres	District	Per 100	District	Per 1,000 acres	District	Per 100
147	23	33	46	189	47	301	64	126	24
145	22	26	45	175	38	302A	51	301	18
301	21	117A	44	206	36	141	37	49	17
302A	21	36A	43	197	35	6	37	147	16
146	21	36	43	26	32	49	36	6	14
141	19	206	42	181	32	302	36	2	13
302	18	47	38	177	29	147	34	155	13
144	18	197	37	186	28	36A	26	141	13
126	17	175	35	207	27	300	25	302A	13
138	17	207	31	205	27	146	25	1	12

Figure 190		Figure 191		Figure 192		Figure 193		Figure 194	
District	Per 100 acres	District	Per 100	District	Per 100	District	Per 100 acres	District	Per 100
328	309	316A	68	6	58	330	126	206	40
330	297	328	67	302	53	329	120	204	39
329	283	306	63	1	53	328	120	207	38
321	270	316	62	2	53	308	114	192	38
308	268	322	62	301	52	321	106	130	37
323	250	329	61	138	49	323	101	316	37
317A	233	315	59	300	49	309	100	90	36
309	226	323	56	302A	48	310	84	13	35
316A	212	321	55	134	48	316A	83	328	34
320	209	325	53	141	48	324	83	317A	34

Figure 195		Figure 196		Figure 197		Figure 198		Figure 199	
District	Per 100 acres	District	Per 100	District	Per 100 acres	District	Per 100 acres	District	Per 100
328	12	141	58	175	143	175	138	189	62
237	12	83	58	32	118	32	116	175	59
322	11	193	57	30	103	30	98	161	57
240	10	209	57	69	95	181	87	32	56
330	9	160	57	53	91	69	87	156	55
17	7	119	57	181	91	53	84	179	52
317A	6	139	57	174	86	174	82	30	51
16	6	173	57	228A	78	176	73	180	51
332	5	7	56	176	76	228A	69	181	50
18A	5	23	56	167	71	167	69	157	49

Selected Statistics

Figure 200		Figure 201		Figure 202		Figure 203		Figure 204	
District	Per 100 acres	District	Per 1,000 acres	District	Per acre	District	Per acre	District	Per 100
32	13	175	125	169	75	169	70	169	69
175	12	32	102	347	42	123	38	123	62
30	10	30	87	123	39	238	35	133	52
181	10	69	78	238	38	347	35	190	45
174	9	53	77	52	31	23	24	347A	43
69	9	181	77	37	29	347A	22	347	38
31	9	174	73	347A	26	133	22	209	37
167	9	176	65	23	25	69	21	163	36
169	8	228A	64	69	23	195	21	52	34
179	8	167	59	133	22	37	18	130	32

Figure 205		Figure 206		Figure 207		Figure 208		Figure 209	
District	Per acre	District	Per acre	District	Per 100 acres	District	Per 100 acres	District	Per 100 acres
347A	15	169	51	168	58	130	353	130	22
123	14	238	26	163	45	164	295	169	8
69	13	347	23	36	37	161	110	181	7
169	10	23	22	91C	26	159	110	303	5
52	9	195	20	173	21	169	102	159	5
133	7	123	15	93	20	181	38	164	5
347	7	37	14	97	15	178	37	79B	3
53	6	336	13	31	14	167	35	69	2
286A	5	133	11	53	13	175	31	160	2
225	5	175	10	160	13	265	20	302A	1

Figure 215		Figure 216		Figure 217		Figure 218		Figure 219	
District	Per acre	District	Per 100	District	Per 100	District	Per 100	District	Per 100
155	19	81	77	3	38	316A	68	125	71
154	16	344A	76	6	38	328	62	192	69
156	15	80	76	4	37	316	60	126	69
153	15	347B	76	301	36	322	57	190	65
228A	14	279A	74	302A	36	306	56	128	65
189	14	80A	73	302	35	315	56	129	63
233	14	344	73	300	34	329	54	194	62
161	14	120	71	321	33	323	51	127	62
197	12	77	71	323	32	321	51	137	60
251	12	350	70	5	32	320	49	134	59

Figure 220		Figure 221		Figure 222		Figure 228		Figure 229	
District	Per 100	District	Per 100	District	Per 100	District	Per 100	District	Per 100
32	27	169	47	155	73	347A	95	323	95
175	23	123	39	154	67	81	92	316A	94
174	18	347	39	197	67	80	92	321	94
31	17	347A	33	228A	67	347B	92	322	93
181	16	52	33	102	67	344A	91	3	91
179	15	238	32	153	65	279A	90	328	90
176	14	133	27	156	63	347	90	329	88
167	13	37	26	235	62	120	89	316	88
53	13	23	25	233	62	344	87	317A	80
28	12	130	22	161	59	80A	87	315	79

Selected Statistics

Figure 230		Figure 231		Figure 232		Figure 233	
District	Per 100	District	Per 100	District	Per 100	District	Per 100
187	82	155	70	175	33	39	33
190	81	197	69	52	27	290	27
126	80	154	67	213A	23	37	26
192	78	233	65	181	23	34A	26
129	77	102	62	225	23	42	26
180	77	156	59	174	21	293	25
128	75	153	58	169	20	38	24
125	74	161	58	176	20	40	22
194	73	235	53	69	20	33	22
56	72	251	52	164	20	31	22

Select Bibliography

References to the regional characteristics of farming in England and Wales are widely scattered throughout a large number of publications, often occurring incidentally in other agricultural studies. Numerous descriptions of the agriculture of different parts of the two countries are to be found in *Agriculture*, especially the several series of farming cameos, in the *Journal of the Royal Agricultural Society*, notably in the Royal Show issues, and in the regional handbooks prepared for the annual meetings of the British Association. The county reports of the Land Utilization Survey describe conditions in the 1930s, and five volumes have so far appeared of the Royal Agricultural Society's post-war series of county agricultural surveys, viz., Kent, by G. H. Garrad (1954); Sussex, by R. H. B. Jesse (1960); Northumberland, by H. C. Pawson (1961); Cheshire, by W. B. Mercer (1963) and Hertfordshire, by H. W. Gardner (1967). Economic studies of many enterprises have been published by the various university departments of agricultural economics, most of which are conveniently summarized in *Digest of Agricultural Economics* (to 1965) and thereafter in *World Agricultural Economics and Rural Sociology Abstracts*. Other publications containing useful descriptions are the *Journal of the British Grassland Society*, *Agricultural Progress*, the *Farm Economist* and the *Journal of the Agricultural Economics Society*. A review of articles and books on agricultural geography which appeared between 1946 and 1964 will be found in J. T. Coppock 'Post-war studies in the geography of British agriculture', *Geographical Review*, 54, 1964, 409–26.

The following list of publications is highly selective, priority being given to those studies in which the distribution of the different crops and livestock is discussed.

General
The contents of these books are relevant to all chapters, but more especially to Chapters IV, V and VII.

Astor, Viscount, and Rowntree, B. S. *British Agriculture*, London, 1938.
Coppock, J. T. *An Agricultural Geography of Great Britain*, London, 1971.
Donaldson, J. G. S. and P. *Farming in Britain Today*, London, 1969.
Duckham, A. N. *The Fabric of Farming*, London, 1957.
Garner, F. H. (ed.) *Modern British Farming Systems*, London, 1972.
Howell, J. Pryse. *An Agricultural Atlas of England and Wales*, Southampton, n.d.

Maxton, J. P. (ed.) *Regional Types of British Agriculture*, London, 1936.
Messer, M. *An Agricultural Atlas of England and Wales*, 2nd edn., Southampton, 1932.
Smith, W. *An Economic Geography of Great Britain*, London, 1948.
Stamp, L. D. *The Land of Britain: its use and misuse*, 3rd edn., London, 1962.
Watson, J. A. S., and More, J. A. *Agriculture*, 11th edn., London, 1962.
Williams, H. T. *Principles for British Agricultural Policy*, London, 1960.
Agricultural Statistics, England and Wales, M.A.F.F., H.M.S.O., annually.
The State of British Agriculture, Oxford, at intervals since 1954.

CHAPTER I

Books

Best, R. H., and Coppock, J. T. *The Changing Use of Land in Britain*, London, 1962, especially chapters 1 and 2.
Butler, J. B. *Profit and Purpose in Farming*, Leeds, 1962.
Monkhouse, F. J., and Wilkinson, H. R. *Maps and Diagrams*, London, 1952.
A Century of Agricultural Statistics, M.A.F.F., H.M.S.O., 1968.

Articles

Coppock, J. T. 'The parish as a geographical/statistical unit', *Tijdschrift voor Economische en Sociale Geografie*, 51, 1960, 317–26.
Duckham, A. N. 'The current agricultural revolution', *Geography*, 44, 1959, 71–8.
Jones, G. T. 'The response of the supply of agricultural products in the United Kingdom to price and other factors', *Farm Economist*, 9, 1961, 537–58, and 10, 1962, 1–28.
Sanders, H. J. 'Balance in British farming', *Advancement of Science*, 16, 1962, 145–54.

CHAPTER II

Books

Bibby, J. S., and Mackney, D. *Land-use Capability Classification*, Tech. Monograph, 1, Soil Survey, 1969.
Reid, I. G. *The Small Farm on Heavy Land*, Wye, 1958.
Taylor, J. A. (ed.) *Weather and Agriculture*, Oxford, 1967.
Agriculture of the Sands Lands, Bulletin No. 163, M.A.F., H.M.S.O., 1954.
The Calculation of Irrigation Need, Technical Bulletin No. 4, M.A.F., H.M.S.O., 1954.
Climatological Atlas of the British Isles, H.M.S.O., 1952.
The Farmer and his Weather, Bulletin No. 165, M.A.F.F., H.M.S.O., 1965.
Irrigation in Great Britain, Natural Resources (Technical) Committee, H.M.S.O., 1962.

Articles

Hogg, W. H. 'The importance of climate in agricultural land classification', *The Classification of Agricultural Land in Britain*, Technical Rept. No. 8, Agricultural Land Service, M.A.F.F., H.M.S.O., 1962, 21–32.
Howe, G. M. 'Climate in relation to crop production', *Agricultural Progress*, 32, 1957, 1–15.

CHAPTER III

Books

Allen, G. R. *Agricultural Marketing Policies*, London, 1959.
Davies, E. T., and Dunsford, W. J. *Some Physical and Economic Considerations of Field Enlargement*, Exeter, 1962.
Denman, D. R., and Stewart, V. F. *Farm Rent*, London, 1959.
National Farm Survey: a summary report, M.A.F., H.M.S.O., 1946.
Scale of Enterprise in Farming, Natural Resources (Technical) Committee, H.M.S.O., 1961.

Articles

Numerous articles in the *Farm Economist*.
Ashton, J., and Cracknell, B. E. 'Agricultural holdings and farm business structure in England and Wales', *Journal of the Agricultural Economics Society*, 14, 1961, 472–98.
Dixey, R. N., and Maunder, A. H. 'Planning again: a study of farm size and layout', *Farm Economist*, 9, 1959, 231–51.
Grigg, D. B. 'Small and large farms in England and Wales', *Geography*, 48, 1963, 268–79.
Kirk, J. H. 'The structure of agricultural marketing in the United Kingdom', *British Journal of Marketing*, 2, 1968, 37–45.
Langdon, A. J. 'Buildings and fixed equipment in agricultural land classification', *The Classification of Agricultural Land in Britain, op. cit.*, 55–60.
Vince, S. W. E. 'Some reflections on the structure and distribution of rural population in England and Wales, 1921–1931', *Transactions of the Institute of British Geographers*, 18, 1952, 53–76.

CHAPTER IV

Books

Britton, D. K. *Cereals in the United Kingdom*, Oxford, 1969.
Sanders, H. G. *An Outline of British Crop Husbandry*, Cambridge, 1959.
Sykes, J. D., and Hardaker, J. B. *The Potato Crop*, Wye, 1962.
The Balance of Arable and Livestock Farming in the United Kingdom, 2nd edn., Milk Marketing Board, Thames Ditton, 1972.
Reports of the Potato Marketing Board, annually, since 1955.
Numerous bulletins of the Ministry of Agriculture, especially: Rotations (85), Potatoes (94) and Sugar-beet Cultivation (153). Also unnumbered booklets on wheat, barley, oats and field beans.

Articles

Britton, D. K. 'The increase in barley growing and its regional pattern', *University of Nottingham School of Agriculture Report*, 1961, 68–74.
Coppock, J. T. 'The changing arable in England and Wales 1875–1956', *Tijdschrift voor Economische en Sociale Geografie*, 50, 1959, 121–30.
Wallace, J. C. 'The development of potato-growing in Lincolnshire', *Journal of the Royal Agricultural Society*, 115, 1954, 60–8.

Select Bibliography

CHAPTER V

Books

Davies, W. *The Grass Crop*, 2nd edn., London, 1960.
Report of the Committee on Grassland Utilisation, Cmd. 547, H.M.S.O., 1958.

Articles

Numerous articles in the *Journal of the British Grassland Society*, e.g., Garner, F. H. 'Ley farming in the Fens', *ibid.*, 1955, 97–105.

CHAPTER VI

Books

Bennett, L. G. *The Horticultural Industry of Middlesex*, Reading, 1952.
Bennett, L. G. *The Diminished Competitive Power of the British Glasshouse Industry because of Mallocation*, Reading, 1963.
Best, R. H., and Gasson, R. M. *The Changing Location of Intensive Crops*, Wye, 1966.
Folley, R. R. W. *Commercial Horticulture in Britain*, Wye, 1960.
Folley, R. R. W., and Hinton, W. L. *British Fruit Farming*, Rept. 64, Cambridge, 1966.
Report of the Committee on Horticultural Marketing, Cmd. 61, H.M.S.O., 1957.
Horticulture in Britain, Part 1, Vegetables, M.A.F.F., H.M.S.O., 1967.
Horticulture in Britain, Part 2, Fruit and Flowers, M.A.F.F., H.M.S.O., 1970.

Articles

Numerous articles in *Scientific Horticulture* on the character of particular areas, especially in volumes between 1932 and 1939.
Cross, P. E. 'The extension of market gardening into agriculture', *Journal of the Royal Society of Arts*, 91, 1943, 310–18.
Furneaux, B. S. 'The soil and the fruit tree', Amos Memorial Lecture, *East Malling Research Station Report, 1949*, 1950.
Hinton, W. L. 'The distribution of horticultural output in England and Wales', *Agriculture*, 67, 1960, 184–8.
O'Connor, J. 'Practical problems in classifying land for horticulture', *The Classification of Agricultural Land in Britain, op. cit.*, 81–90.
Pocock, D. C. D. 'England's diminished hop acreage', *Geography*, 44, 1959, 14–21.
Round, K. M. 'Celery growing in the Fens', *Agriculture*, 67, 1960, 189–91.

CHAPTER VII

Books

Allanson, G. *Kent or Romney Marsh Sheep*, Wye, 1961.
Attwood, E. A., and Evans, H. G. *The Economics of Hill Farming*, Cardiff, 1961.

259

Coles, R. *Development of the Poultry Industry in England and Wales, 1945–1959*, London, 1960.

Costs and Efficiency in Milk Production 1968–1969, M.A.F.F., H.M.S.O., 1972.

Dairy Herd Census 1970, Milk Marketing Board, 1971.

Disease, Wastage and Husbandry in the British Dairy Herd, M.A.F.F., H.M.S.O., 1960.

The Pig Industry in Great Britain in 1960, Pig Industry Development Authority, 1960.

Reports of the Production Division, Milk Marketing Board, annually.

The Sheep Industry in Britain, Natural Resources (Technical) Committee, H.M.S.O., 1958.

Enterprise studies from university agricultural economics departments, particularly the series *Economic studies of sheep farming in Wales* (Aberystwyth) and *Some economic aspects of the sheep industry in the west of England* (Bristol).

The following bulletins of the Ministry of Agriculture: Rations for livestock (48), Sheep breeding and management (166), Cattle of Britain (167) and Beef production (178).

Articles

Barnes, F. A. 'The evolution of the salient pattern of milk production and distribution in England and Wales', *Transactions of the Institute of British Geographers*, 25, 1958, 167–95.

Bellis, D. B. 'Pig farming in the United Kingdom', *Journal of the Royal Agricultural Society*, 129, 1968, 24–42.

Bowman, J. C. 'The egg production industry in the United Kingdom', in Bunting, A. H. (ed.). *Change in Agriculture*, London, 1970, 263–75.

Coles, R. 'The poultry industry today', *Association of Agriculture Journal*, 8, 1966, 2–9.

Hart, J. F. 'The changing distribution of sheep in Britain', *Economic Geography*, 32, 1956, 260–74.

Long, W. H. 'The place of beef production in British farming', *Journal of the Agricultural Economics Society*, 9, 1950-2, 4–13.

Robinson, J. F. 'Breed adaptation and environment in sheep', *Journal of the Royal Agricultural Society*, 120, 1959, 20–7.

Simpson, E. S. 'Milk production in England and Wales', *Geographical Review*, 49, 1959, 95–111.

Strauss, E. 'The structure of the English milk industry', *Journal of the Royal Statistical Society*, Series A, 123, 1960, 140–65.

CHAPTER VIII

Books

Farm Classification in England and Wales 1969–1970, M.A.F.F., H.M.S.O., 1971.

Farm Income Series, M.A.F.F., H.M.S.O., annually.

Type of Farming Maps in England and Wales, M.A.F.F., H.M.S.O., 1969.

Articles

Britton, D. K., and Ingersent, K. A. 'Trends in concentration in British agriculture', *Journal of Agricultural Economics*, 16, 1963–4, 26–52.

Coppock, J. T. 'Crop, livestock, and enterprise combinations in England and Wales', *Economic Geography*, 40, 1964, 65–81.

Hirsch, G. P. 'Labour requirements and availability in British agriculture', *Farm Economist*, 9, 1961, 518–25.
Napolitan, L., and Brown, C. J. 'A type of farming classification of agricultural holdings in England and Wales according to enterprise patterns', *Journal of Agricultural Economics*, 15, 1962–3, 595–616.

CHAPTER IX

Book

Carlyle, W. J. *Some Aspects of the Geography of Livestock Movement in Scotland*, unpublished Ph.D. thesis, University of Edinburgh, 1970.

Article

Coppock, J. T. 'The geography of agriculture', *Journal of Agricultural Economics*, 1968–9, 153–69.

Index

Figures in parentheses are the page numbers of relevant illustrations

262

Index

Index

Wisbech, 69, 155, 157
Worcestershire, 23, 60, 63, 95, 130,
  139, 147, 157, 158, 159, 161, 163,
  164, 209, 218
workers, agricultural, 60–2 (61)
World Agricultural Census, 24
World War, First, 21, 51, 202;

Second, 21, 62, 86, 87, 94, 104,
  113, 117, 121, 130, 141, 145, 148,
  151, 153, 169, 177, 192, 198, 200,
  202, 203
Worthing, 150

Yarmouth, 69

yields, 25; see also *major crops and
  grass*
Yorkshire, 90, 93, 99, 100, 103, 190,
  196, 200, 203
Yorkshire, Wolds, 31, 42, 124, 218
York, Vale of, 31, 44, 89, 95, 102,
  139, 166, 182, 184, 187

267